The Hamlyn Guide to

Weather Forecasting

S. Dunlop F. Wilson

HAMLYN
London · New York · Sydney · Toronto

Acknowledgements

The figure drawings are by the Hayward Art Group.

The photographs on pages 119 top, 124 top, 124 bottom and 125 are Crown copyright reproduced by permission of the Controller of Her Majesty's Stationery Office; Colour Library International, London 21, 52-53; Peter Crump, Beaconsfield 35 top, 35 bottom, 44, 90; Decca Radar Ltd., London 119 bottom; Storm Dunlop, East Wittering 47 top, 51 bottom, 57 top; European Space Agency (ESA), Darmstadt 126; Karel Feuerstein, Ruislip 1, 51 top, 59; Keystone Press Agency, London 15, 16; Frank Lane, Pinner 29; Wilson A Bentley US Weather Bureau 48 top; Peter Davey 37 bottom, 100-101; Eichhorn/Zingel 104; W T Miller 22, M Nimmo 14, 38-39, 43 top; NOAA 105; Mandal Ranjit 85; R S Virdee 28; Peter Loughran, High Wycombe 49, 55 top, 99 top, 146; Michael Maunder, 55 bottom, 75, 107; Meteorological Office, Bracknell 86-87; Captain S A Greenaway 77 bottom; NASA, Washington 6, 71 bottom, 110-111; NOAA, Washington 23, 72; R K Pilsbury FRPS, F R Met S, Totland 17, 19 top, 27, 37 top, 41, 42, 43 bottom, 45 top, 45 bottom, 47 bottom, 53, 54, 57 bottom, 61, 62, 63, 64, 71 top, 73 top, 77 top, 81, 83, 87, 89, 91, 94-95 bottom, 97, 98 left, 98 right, 99 bottom left, 99 bottom right, 103, 143; RIDA, Norbiton – Alex Maltman 48 bottom; R T J Moody 19 bottom, 68-69, 94-95 top; R Towse 78-79; Shell UK, London 73 bottom; University of Dundee 127, 128 top, 128 bottom, 130, 131.

Published by
The Hamlyn Publishing Group Limited
London · New York · Sydney · Toronto
Astronaut House, Feltham, Middlesex, England

ISBN 0 600 39024 1 (paperback)
ISBN 0 600 35619 1 (cased edition)

Printed in Spain by Printer Industria Gráfica SA
Sant Vicenç dels Horts Barcelona Depósito Legal B. 23075-1982

Contents

Introduction 4
The global circulation of the atmosphere 6
Seasonal effects 10
Pressure 14
Winds 18
Air masses 20
Anticyclones and depressions 24
Atmospheric layers and temperatures 28
The formation of clouds 30
Cumulus and cumulonimbus 34
Stratocumulus 38
Altostratus and altocumulus 42
Cirrostratus, cirrocumulus and cirrus 44
Clouds and precipitation 46
Sky and cloud colour 50
Optical phenomena 54
Rainbows and fogbows 58
Mountains, hills and air waves 60
Hills, clouds and rain 62
Valleys 66
Coastal regions 70
Fog 74
Frost 78
Soil and vegetation 82
Towns, cities and pollution 84
The depression sequence 88
Showers and hail 98
Thunderstorms 102
Tornadoes, waterspouts and other whirlwinds 106
Tropical storms 108
North American weather 112
European and Mediterranean weather 114
The weather of Australia and New Zealand 116
Professional forecasting 118
Understanding forecasting language 120
Television forecasts and isobaric charts 121
Interpreting satellite photographs 126
Radio forecasts 132
Telephone and newspaper forecasts 134
Making use of forecasts 134
Longer range forecasting 136
Records and instruments 140
Forecasting likely maximum temperatures 148
Forecasting overnight temperatures and precipitation 150
Making a forecast 152
Glossary 153
Bibliography 154
Weather map symbols 155
Index 158

Introduction

It is the continual variability of the weather – and the problems of forecasting it properly – which causes it to be of interest and concern to everyone. There are very few regions of the Earth where conditions are completely predictable for most of the time, and even there sudden extreme events can occur. It is the temperate zones in the middle latitudes of both the northern and southern hemispheres which undergo the most frequent changes, so that the conditions encountered there form a major part of this book. The weather and climate of the polar and equatorial regions are not covered to any great extent.

The general aim is to enable the reader firstly, to recognize the form of weather, the clouds and other phenomena which are present at any time; secondly, to understand how they have been produced; and finally, to move on to forecasting what is likely to occur. The arrangement is such that the book may be read in a logical order, progressing from global aspects of weather and general processes (*pp. 6-29*), clouds and other phenomena (*pp. 30-59*), through to local conditions (*pp. 60-87*). This is followed by details of the important sequences of weather conditions found in low pressure systems, showers and thunderstorms, and more extreme storms (*pp. 88-111*). A brief description of general weather patterns over North America, Europe and the Mediterranean, and Australia and New Zealand, gives some idea of conditions encountered in those parts of the world (*pp. 112-117*). Some aspects of professional forecasting are discussed, together with ways in which these, and satellite photographs, may be interpreted (*pp. 118-139*). Finally records and instruments are dealt with, as well as the means of making certain specific forecasts.

In a complex subject such as meteorology it is difficult to isolate one aspect from all the others, so that some conditions may be mentioned in several places. The major discussion will be found where the subject is given in bold type. Extensive cross-references are given, and the index and contents list should be freely used. Although this book should offer something for nearly everyone, it is naturally not possible for it to go into detail in more specialized areas, such as aviation meteorology, for example, but some guidance to more detailed explanations will be found in the bibliography.

All measurements are primarily given in terms of the metric (SI) system, in accordance with the world-wide usage which this now has, and taking into account the growing tendency for books and periodicals published in the U.S.A. (and elsewhere) to use these units. However non-metric equivalents are also given, but these are approximate conversions and are not intended to be over-precise. It is strongly recommended that anyone undertaking practical meteorological work should, for example, purchase thermometers graduated in degrees Celsius, and use metric units throughout.

It is just not possible to discuss or to illustrate clouds in very great detail in this book, as due to the many different classified varieties and also because of their continual change of form, a great deal of space would be required to even begin to cover the subject. Even such major reference works as the *International Cloud Atlas* published by the World Meteorological Organization can only show a small proportion of the endless variations which make them such a fascinating study. However, in selecting the cloud illustrations an effort has been made to show some representative forms which will assist with general identification, rather than concentrating upon unusual types, although some of these are included.

As clouds may form at many levels, and several types may be present at once, their identification is sometimes regarded as being difficult. It is true that it is not always easy, but anyone who is prepared to spend a little time watching the sky will find, with increasing experience, that most types can be identified quite easily and this will certainly be enough for most practical purposes.

A somewhat similar situation occurs in making a forecast, in that at first it appears difficult to do, if not quite impossible. However, once again with practice, and perhaps a small degree of patience, a very reasonable degree of success can be obtained. It is important to remember that all forecasts, professional or amateur, have their own limitations, but that local knowledge will help to make them more accurate. Guidance is given into the interpretation of national and regional forecasts which have been professionally prepared, to take local conditions into account.

The art of weather forecasting exerts a fascination all of its own. It is hoped that this book will help readers to find this out for themselves, as well as encouraging their interest in the signs shown by the ever-changing sky.

The global circulation of the atmosphere

It is quite possible to forecast the weather without having any insight into the physics of the atmosphere, simply by keeping an eye on the changes taking place in the sky and learning by rote what they mean. But an amateur forecaster will find the process easier and more interesting, and in the long run probably more reliable, if he has some basic information about why the earth's atmosphere behaves in the way it does.

The thin layer of the atmosphere enveloping our planet Earth is held to it by the force of gravity, which prevents it from being lost to space, and also restricts its vertical motions. Since the Sun delivers uneven heating across the face of the Earth the equator is heated more than the poles; the air in contact with the Earth's surface in the equatorial region is heated, and as it is heated it rises into the upper atmosphere and

Extensive cloud coverage of the southern hemisphere is shown in this Apollo 17 photograph taken in December 1972. Depressions over the seas around the ice-covered Antarctic continent contrast sharply with cloud-free northern Africa and Arabia.

flows poleward. If nothing else were to happen a simple circulation of hot air rising over the equator and cold air sinking over the poles would be established. However, because the world spins from west to east, rotational forces deflect winds to the right in the northern hemisphere and to the left in the southern, causing the entire circulation pattern to be fare more complex. Distinct circulation cells are established over the tropical and polar regions of both hemispheres, causing the surface winds there to be, on average, easterlies, that is to blow towards the west. But over the middle latitudes the surface winds tend to be westerlies (*Fig. 1*).

It is the thermally-driven vertical circulation of the air (or convection) that is responsible for the low surface pressures at low latitudes or over heated regions (*Fig. 2a*), and also for the high pressure over the poles and similarly cooled areas of land in winter (*Fig. 2b*). However, in the middle latitudes where the most disturbed

Fig. 1 A simplified representation of global air circulation and prevailing surface winds. The general pattern is the same for both hemispheres apart from seasonal effects and variations due to the distribution of land and sea areas.

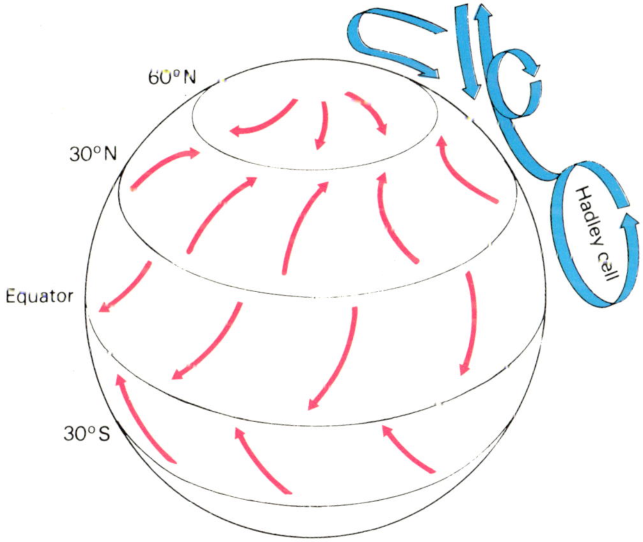

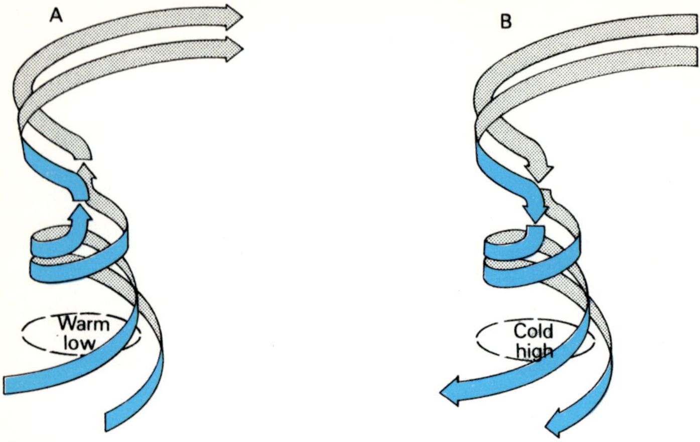

Fig. 2 *above* The basic convection processes. *Left* heated air rises and expands, causing a warm surface low pressure area, while (*right*) cold, dense air subsides, giving rise to a cold high.

Fig. 3 *right* High-speed airflow at height encircles the world at several latitudes in a series of waves. The precise location of jet streams varies considerably as does their velocity.

Fig. 4 *below* Convergence at height (*left*) causes air to subside and grow warmer, producing a warm high at the surface with diverging airflow. Divergence at height and convergence at the surface are found in a cold low (*right*).

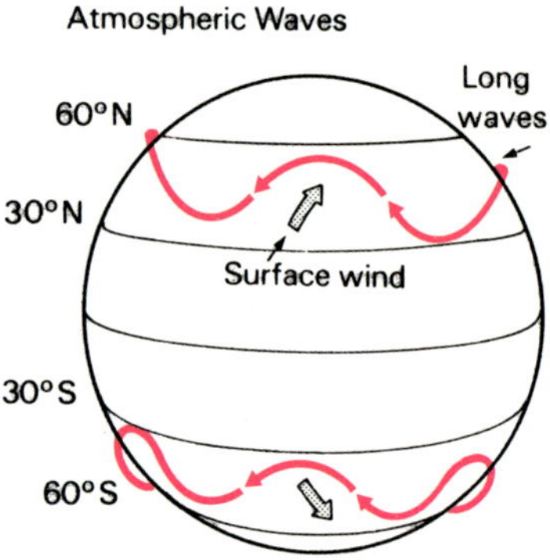

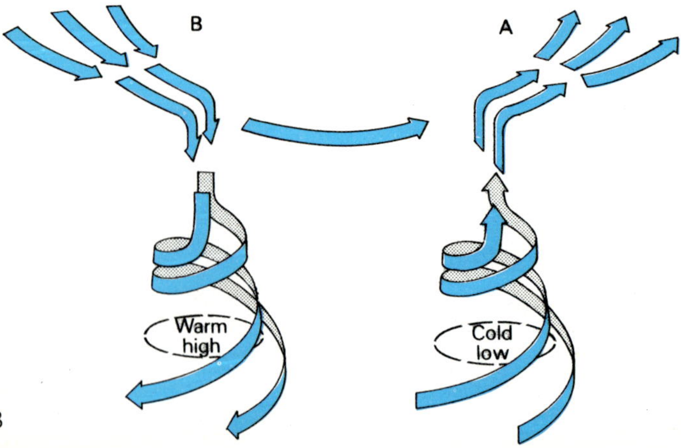

weather patterns are found, the basic vertical motions of the air have predominantly mechanical causes. The full explanation is complex, but essentially the combined effect of the uneven solar heating, the Earth's rotation and the high mountain ranges of the world is to produce a series of four or five 'long waves' in the upper atmosphere of each hemisphere between altitudes of about 6–12 km (4–8 miles). These waves encircle the Earth with relatively high-speed winds along a track that wanders slowly over the middle latitudes (*Fig. 3*). The air moves around the world through these waves as it acts to even out the heat imbalance.

On their own these fundamental waves would lead to fairly predictable weather patterns. However, localized strong thermal gradients in the air introduce subtle small perturbations, or secondary waves, which themselvos run around the world through the long waves, and these smaller waves greatly complicate the picture.

All these upper air waves are by no means symmetrical and as the wind streams accelerate and decelerate around their crests and troughs, the air flow is forced to converge and diverge. However, converging upper air cannot go on accumulating indefinitely, so the air below begins to sink out of the way. At low levels it becomes compressed and consequently the atmospheric pressure, which can be measured by a barometer, rises. The subsiding air is warmed as it falls, so any clouds tend to thin and evaporate, allowing more solar heating by day and greater torrestrial cooling by night. The surface air flows outwards and the weather becomes settled (*Fig. 4a*).

Over regions where the upper air diverges the air below begins to rise up to take its place, relieving the air pressure, and the barometer falls (*Fig. 4b*). Surface winds converge inwards, bringing with them contrasting warm and cool air masses. But these air masses do not readily mix and instead the denser cold air drives underneath the lighter warm air, which is forced to ride upwards, eventually cooling and condensing into clouds that grow high enough to produce rain or drizzle, hail or snow. The weather in the region has become unsettled.

This up and down movement in the atmosphere as the upper winds weave their way around the world through the long fundamental waves and also around the sharp crests and troughs of the smaller migratory waves, is the essence of the disturbed weather of the middle latitudes. Sometimes those upper waves become blocked causing long spells of unchanging weather; sometimes they reverse direction, bringing unseasonal extremes of weather to the underlying regions.

So what we call the weather is really just the atmosphere in motion. The total volume of the atmosphere is so enormous, and the factors influencing its movement so varied, that there is always the potential for sudden dramatic changes. Weather operates in such a complex way that it is not surprising that the professional weathermen cannot always forecast the changes that will occur at any given time.

Seasonal effects

The Earth's axis around which it rotates is not vertical to the plane of its orbit about the Sun. As a result during the half-year from March 21 to September 21 the northern hemisphere is tilted towards the Sun and the southern away from it; for the remaining half-year the positions are reversed. This variation produces the different seasons and the corresponding alteration in the weather patterns across the globe.

The major consequence of this tilt is that the Sun delivers more heat in the summer to the surface of the land and sea than in the winter, and in mid-latitudes for example, summer heating is about six times that in

Fig. 5 The average surface temperature distribution for January (*left*) and July. The shift of the hottest regions south and north of the Equator can be clearly seen.

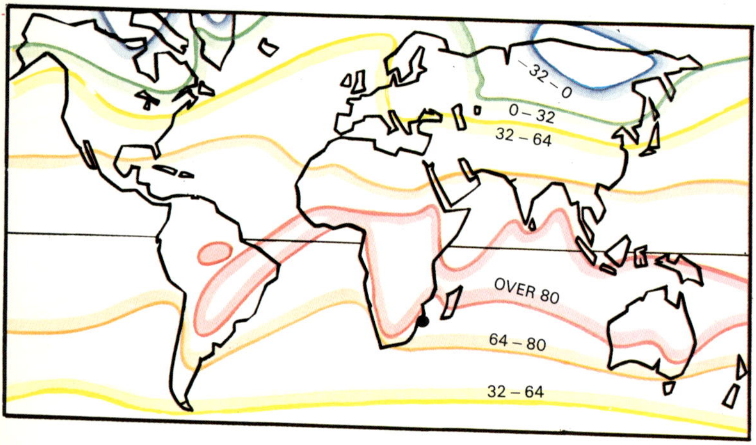

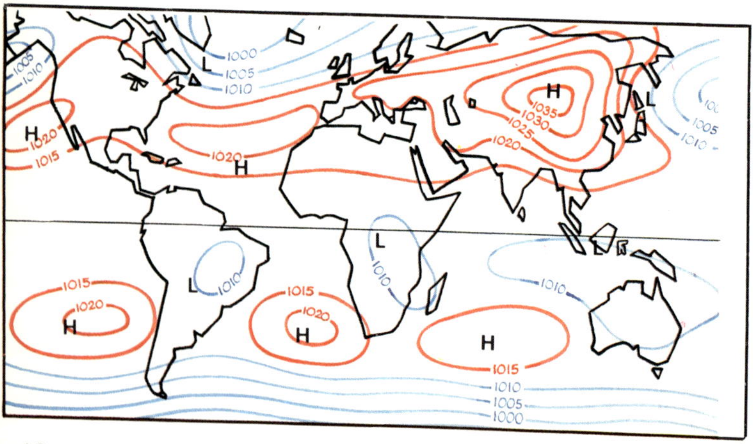

winter. In the tropics the yearly variation is less, but it is still considerable. Within the Arctic and Antarctic Circles, of course, the long dark winters are responsible for considerably increasing the size of the large masses of cold air which exists over those regions throughout the year and which exert great influence over the general circulation.

The variation in global temperatures is shown in *Fig. 5*, where the months of January and July are compared. It is noticeable how the highest temperatures are strongly concentrated in the northern hemisphere in July, and only the interior of Australia has strictly comparable temperatures in January. The ranges of temperature at certain selected locations are also shown.

Fig. 6 The average surface pressure for January (*left*) and July. The warm lows roughly corresponding to the hottest regions can be seen, although the Azores High is important during the northern summer.

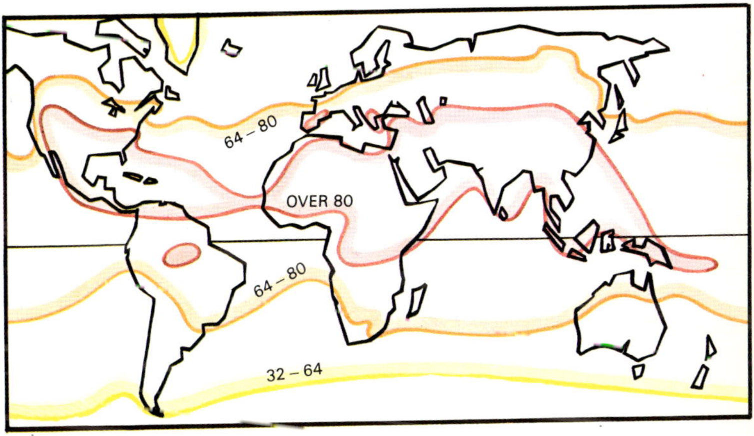

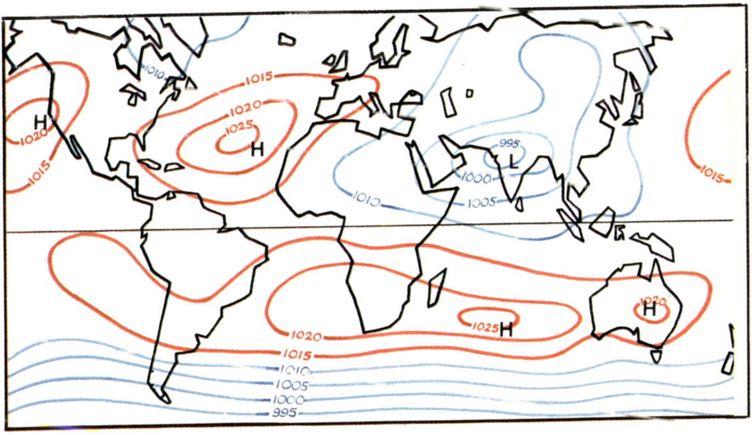

The alteration in surface pressure shown in *Fig. 6* follows the seasonal variation in temperature. In the southern hemisphere the high pressure zone located at approximately 25–30° south is present all the year, although strengthening considerably in summer. In contrast, the northern hemisphere could be said to be dominated by the cold Siberian and North American highs in winter, and by the warm subtropical high pressure areas and the Asian warm low which is present in summer.

As a result of the changes in pressure with the season there are, of course, alterations in the general pattern of surface winds produced by the global circulation of air described earlier. The average wind directions are shown in *Figs. 7* and *8* for January and July. The most notable change is that of the monsoon winds, particularly those of south-

Fig. 7 The predominant air masses and winds of the Northern Hemisphere during winter (*left*) and summer. The meaning of the abbreviations for the air masses is explained on p. 20.

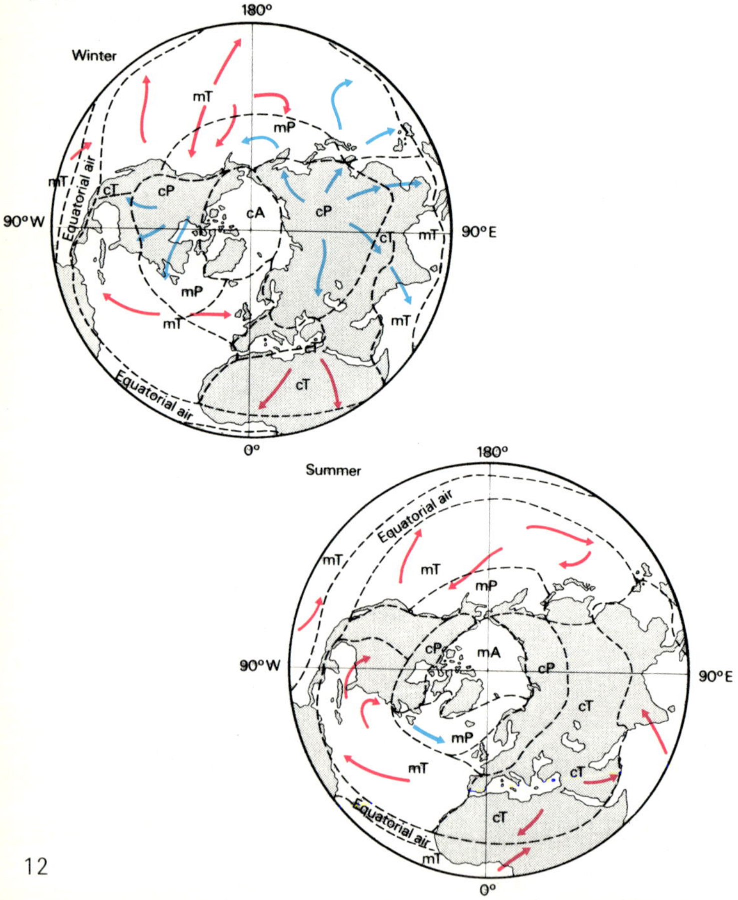

eastern Asia and India, where the north-easterly airflow prevailing in winter with its generally dry conditions, suddenly changes to the very warm moist south-westerly monsoon. It is this south-west monsoon which is of such vital importance to agriculture in the region as it is responsible for the greatest part of the annual rainfall on which the crops are so dependent.

As a result of the pressure and wind patterns, various regions of the world may be described as generally lying under the influence of particular air masses and *Figs. 7* and *8* show these. It should be emphasized that these are only very approximate representations of the overall pattern and that those regions on or near the boundaries of the various air masses, particularly the polar and tropical ones, experience very variable conditions.

Fig. 8 The summer and winter air masses and circulation of the Southern Hemisphere. Note how much simpler the patterns are compared to the North, where the land areas have great effects.

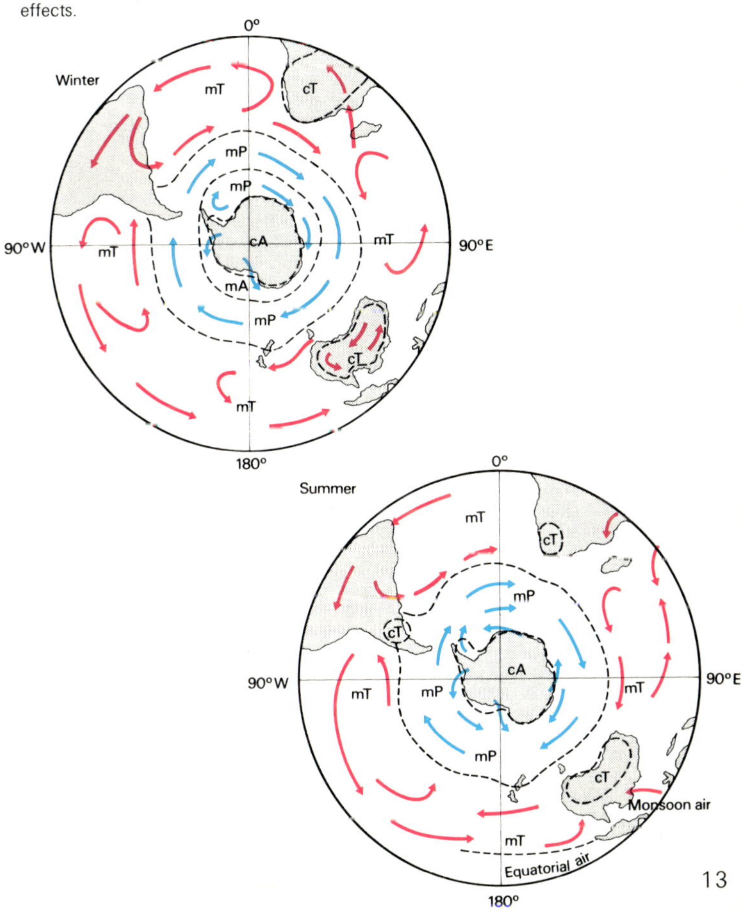

Pressure

The most important single factor to consider when determining what sort of weather the present and the immediate future hold is the atmospheric pressure – this exerts an overriding influence. Atmospheric pressure is simply the weight of the enormous column of air that rests on a given unit area, and it is measured by means of a barometer.

Barometers are usually calibrated in **millibars** (mb), or thousandths of a bar, which is defined in terms of metric (SI) units. This gives rise to the fact that the average atmospheric pressure at the surface of the earth has the apparently strange value of 1013·2 mb.

Pressure, of course, varies with height in the atmosphere and the relationship is shown in *Fig. 14*. In order to enable pressure maps to be drawn which represent the true state at any time, corrections are made for the height of the observing stations, and in practice their barometers are adjusted to show the pressure at mean sea-level. This

Typical spring or summer cumulus clouds formed by convection over land warmed by the Sun. These clouds build up during the morning and afternoon, but disperse in the evening and overnight.

Continuous heavy rain associated with a low pressure system causing problems at a tennis tournament. Unlike areas with continental climates, maritime regions – in this case England – may suffer such conditions even in mid-summer.

reduction to sea-level can be difficult, even when as many factors as possible are taken into account, so that some uncertainty remains. In mountainous regions, pressure charts can only be an approximation to the truth. The usual sign of very high pressure is a dry spell, with mostly clear skies and little wind. Very low pressure, by contrast, usually produces dull, wet and windy weather. However, this is only the beginning of what barometer readings can tell us. Barometer manufacturers have taken advantage of the obvious correlation between certain weather conditions and extremes of atmospheric pressure to produce instruments which interpret the pressure in a series of misleading specific stages, such as 'stormy – rain – change – fair – very dry'. At the ends of the scale these may be fairly accurate, but the labels in between may not be. The level of the air pressure does not, on its own, correspond to any definite weather description.

In fact, the way that pressure in changing is just as significant as where the pointer stands at any given moment. Assuming you have a barometer, then the pressure tendency can be found by noting the present reading and returning to read it again in one or two hours. High pressure that is also rising rapidly brings the best type of weather. Pressure will continue to rise so long as the upper air convergence goes on exceeding the surface air divergence. In these circumstances the upper air is slowly sinking, that is, being compressed. Just as expansion causes any gas to cool, the converse also holds, therefore the slowly sinking air is warmed. The ice crystals and water drops that make up the clouds evaporate and thin, or dissolve completely. The higher the pressure and the stronger the rise the more cloud-free the weather becomes.

Often, though, pressure is only moderately high and rising slowly.

Low pressure systems frequently bring high winds and consequent high seas. The increase in water height caused by the lowered pressure may help to produce flooding of low-lying coasts.

Then bubbles of air, usually warmed by contact with the land in summer or the seas in winter, can still convect upwards like hot air balloons, expanding and eventually cooling to saturation. If saturation point is reached before they meet the sinking air aloft, shallow bubbly clouds form and spread out where they meet the warm upper sinking air to form broken or complete sheets of shallow cloud. These clouds are usually not deep enough to produce rain or substantially reduce the amount of direct sunlight, so the day will still be fairly bright.

On the other hand, low pressure that is also falling rapidly is a sign of the worst type of weather. Pressure will continue to fall so long as the upper air divergence exceeds the low level convergence. The low level convergence brings into conflict contrasting cool and warm air masses, into a region of generally ascending air. Normal warm air rising into cooler air above also assists this process. The uplift of the warm air mass is caused by the denser cool air mass moving beneath the warmer air, forcing it to ride upwards. The subsequent cooling to saturation of the warmer air, and continued cooling by uplift to even colder temperatures, is on such a scale as to produce huge towers of rain clouds in vast banks covering large areas. Heavy and prolonged rain results, and the clouds are so deep that they reflect almost all the sunshine back into space, making the day dull or even very dark.

The upcurrents in a low pressure area are very much stronger than the downdraught in a high pressure system. Strong winds are therefore another distinctive feature of low pressure areas, bringing with them violent conditions and storms. The lower the pressure and the stronger the fall the more intense the storm.

If, though, the pressure is only slightly low (i.e. around 1000 mb) and falls slowly, then the low level convergence is much less vigorous. Clouds of great depth may still form, but even when they are numerous they are still usually separate in the sky. These clouds are loosely labelled 'potential shower clouds'.

In order to maintain the essential balance between the amounts of rising and falling air over the world, high pressure systems (**anti-cyclones** or **highs**) cover much larger areas of the Earth than low pressure systems (**cyclones, depressions** or **lows**), and their extensive nature means that they are far less mobile than the smaller depressions. So the stormy weather associated with lows can be here today and gone tomorrow, whereas settled weather is longer-lasting.

The range of pressures encountered varies from below 940 mb in the most intense depressions to over 1050 mb in the strongest anti-cyclones. Normally however, especially in temperate latitudes and where the region is not affected by tropical storms, the variation will be much less than these extremes.

Large cumulus clouds and heavy showers of rain associated with the cold front of a depression. Such a belt of rain may pass in a few hours to be followed by more settled weather.

Winds

Meteorologists plot the distribution of pressure on maps known as isobaric charts. Although these will be discussed in detail later, some consideration is useful here. Pressures are shown by the use of isobars, frequently spaced at intervals of 4 millibars on general charts. Such lines of equal pressure will obviously be closed around high and low pressure centres, which will thus show on the charts as roughly circular or elliptical areas. Air tries to flow directly from a high pressure area to one of low pressure, but because of the rotational forces mentioned earlier (*p. 7*) – the Coriolis effect – and friction, the motion is altered so that in the northern hemisphere surface winds circulate clockwise about high pressure centres (*Fig. 9*). Around depressions the motion is anticlockwise. The closer the isobars, the greater the wind strength. Rather than flowing along the isobars, surface winds are at a small angle to them – very nearly parallel, in fact.

A practical, and very useful, consequence of this is that from the wind direction it is possible to determine the position of high and low pressure. This, the Buys-Ballot law (named after the Dutch scientist), states that in the northern hemisphere outside the equatorial regions an observer with his back to the wind has *low* pressure on his *left*.

The angle which the winds make to the isobars depends upon the surface friction (*p. 60*) and will be least (about 10°–20°) over the sea and greater (about 25°–35°) over the land. This frictional effect lessens with height and so gives rise to a shift in the wind direction. At 500-1000 metres (1640-3280 ft) the wind is flowing freely along the isobars, and this higher wind (actually at about the height of the low clouds) gives an even better indication (under the Buys-Ballot rule) of the directions of low and high pressure.

The air above any particular spot may be expected to consist of more than one layer of differing temperature and humidity. The pressure distributions and winds will also vary, generally becoming simpler with height, with the high-speed flows mentioned earlier also obeying the Buys-Ballot law in their oscillations around the world. Where the flow in the upper waves is particularly strong it is described as forming a **jet stream** and as we have seen influences the formation of high and low pressure areas at the surface.

Fig. 9 Idealized patterns of air flow in the Northern Hemisphere high (*left*) and low pressure areas, with isobars shown as circles. In the Southern Hemisphere the circulations are anticlockwise and clockwise around highs and lows respectively.

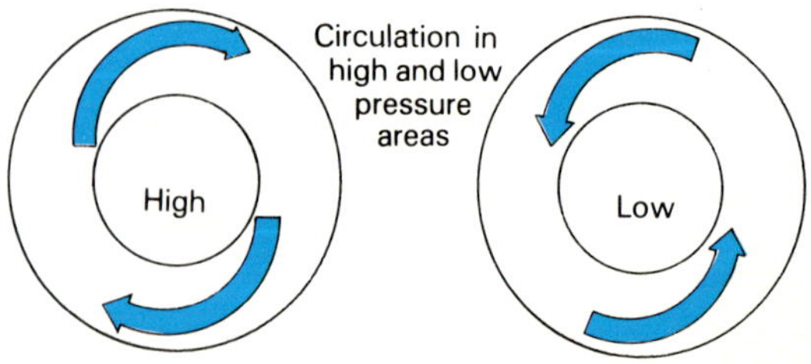

These jet-stream clouds at a height of about 9 km (30 000 ft) have strong parallel banding along the wind direction. Observations showed that the wind speeds were about 260 kph (140 knots or 160 mph).

As clouds may occur in all the layers we are considering, their movement gives a very useful indicator of the weather which is in store. The highest, the cirrus clouds (*p. 44*), particularly jet stream cirrus, give early warning of the approach of a low pressure system, while middle clouds (*p. 33*) show when it is still nearer.

It is of course the variation in wind direction in the various layers which accounts for the (false) statements sometimes heard about clouds which approached 'against the wind'. Quite apart from the effects discussed here, many purely local factors can cause the surface wind to differ greatly from that at higher levels. Even at sea, a squall bearing 10°–20° to the right of the surface wind direction poses more danger to a sailor than one directly to windward, as they both move in accordance with the higher-level wind.

Winds can play a major part in shaping the Earth's surface. These high sand dunes at Kerzaz in Algeria are just one example of how loose materials may be transported for considerable distances.

Air masses

An **air mass** is a large volume of air – large enough to cover an area of thousands of square kilometres or miles – that moves in a body and within which temperature and humidity remain fairly constant. The prevailing air mass therefore has an overriding influence on the type of weather to be expected over extensive land areas, although at a local level the broad pattern will be modified by many topographical features such as hills, towns, stretches of water and so on (the effects of such local features are discussed in detail later, see *pp. 60–73*).

Air masses are classified primarily according to two factors: the first is their origin; the second is the nature of the terrain over which they have travelled. The origin is important because the temperature of an air mass depends directly on its source: air masses which are **polar** (designated P) or **arctic** (A) in origin are cold; those which are **tropical** (T) or **equatorial** (E) in origin are warm.

Care should be taken to distinguish between the terms 'polar' and 'arctic', which may cause confusion. ('Arctic' is usually taken to imply 'antarctic' as well in general discussions, but when the latter air mass must be distinguished the letters AA are used.) Due to the way in which the study of meteorology developed, the polar air masses had been so named before it was realised that distinct air masses (arctic) from even higher latitudes were of very considerable importance. Very approximately, arctic or antarctic air originates within the Arctic and Antarctic Circles, or more specifically, is the air over the two ice-caps.

The temperature of an air mass is determined by its origin, but its second important quality, its humidity, depends on whether it has travelled over land or over the sea, or both. If its track is over land it is called **continental** (symbol c) and is dry; if over the sea it is called **maritime** (m) and is moist. We may combine these factors to describe the characteristics of the air masses, but equatorial air is always maritime in nature (mE), and will not be discussed here. Continental arctic (cA) air as such is only of major significance over the polar ice caps, and can only be readily distinguished from continental polar air (cP) by the temperature at height, so it also will not be considered in detail. However, in the Northern Hemisphere it can on occasions sweep down over Canada and the central U.S.A., and parts of Eurasia, bringing dry, cold, and in winter, bitterly cold, weather.

We find that there are five common types of air mass that frequently affect mid-latitude regions: maritime arctic, mA; continental polar, cP; maritime polar, mP; continental tropical, cT; maritime tropical, mT.

Modifications to the broad characteristics of air masses are common. Apart from the effects of local land features, the air mass itself

Fig. 10 The lobed pattern of northern air masses in winter. The tongues of cold air shift their position around the world in accordance with the upper atmosphere waves.

Fig. 11 The representation of fronts on meteorological charts. The cusps are shown on the side towards which the front is moving.

Sub-arctic conditions on the shore of the Bering Sea near Nome in Alaska. Cold air building up above such regions and the high Arctic exerts considerable influence over the weather further south.

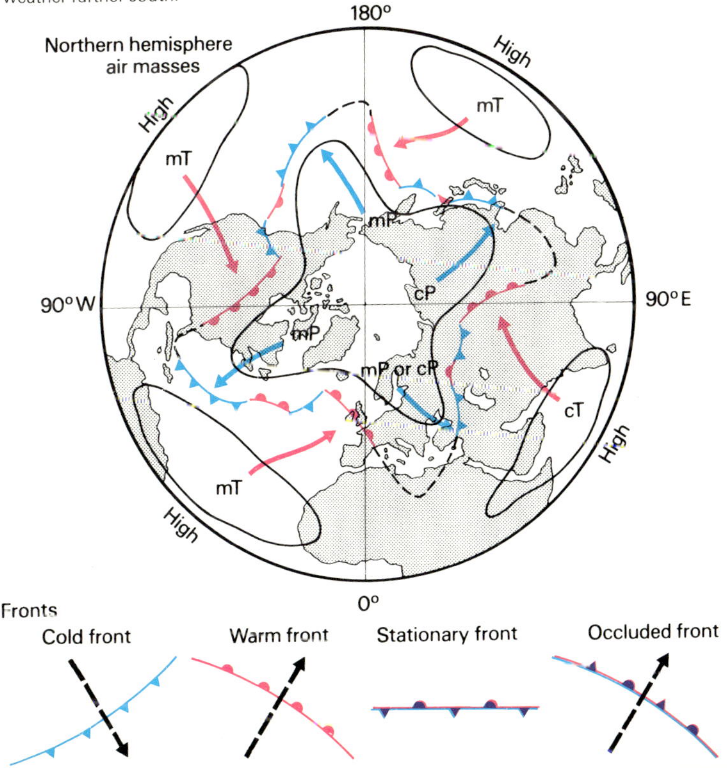

Small clouds forming over the Kalahari Desert. The restricted amount of moisture in the air means that they will not grow much larger nor give rise to showers.

may approach in a roundabout way so that its original character, typical of its source region, is considerably altered. Furthermore, atmospheric pressure (*pp. 14–17*) has an influence which can be so powerful that it sometimes obscures the qualities of the air mass itself. However, these are the general characteristics of the various types:

Maritime arctic air, mA, is only found with reasonable frequency over Western Europe. Cold in summer, very cold in winter, it has travelled for quite a long distance over a warmer sea, and has warmed, picked up moisture, and become unstable (*p. 114*) thus producing frequent showers which are usually of snow in winter.

Continental polar air, cP, shows considerable differences in winter and summer. In winter the cooling over the centres of the North American and Eurasian land masses produces very cold, very dry, stable air, resembling continental arctic air. In summer the source regions are subject to quite high temperatures during the long days, and the cP air may begin as fairly cool and dry, but travelling south it absorbs moisture and becomes more unstable. It should be noted that cP air is not found in the Southern Hemisphere as there are no land masses south of 50° suitable for its generation. The air flow is therefore always of maritime polar nature.

Maritime polar air, mP, frequently affects many regions of the world, such as the western and eastern seaboards of North America, Western Europe, southern Australia, and New Zealand. It is also cool when it

A cold, high pressure air mass over northern Canada is sending a wedge of polar air down as far as the Dallas/Fort Worth area of Texas. The wedge is bounded on the west by a warm front running from Wyoming to Amarillo, Texas, ahead of warm air from a high-pressure area over the south-western states; in the south-east the cold front runs up to a low centred over Kentucky, with another warm high over Florida. The polar air streaming over the East Coast picks up considerable moisture above the warm Gulf Stream.

begins but warms and picks up moisture during its passage over the warm ocean, becoming unstable and frequently bringing clear, cold, and showery weather.

Continental tropical air, cT, builds up in air masses over the desert regions of the world and is hot, dry, and unstable, although the lack of moisture means that few clouds are produced. The sources over North Africa and in the interior of Australia tend to be present all the year, but the others decline in winter.

Maritime tropical air, mT, is warm and moist (in summer very warm and very moist) and deposits large quantities of rain (or snow, in high latitudes), especially on windward coasts.

The boundaries between contrasting air masses are known as **fronts.** They are usually not only regions of contrasting temperatures, but also of great **wind shear,** that is, where the wind velocity changes sharply over a very small distance. The boundary between the polar easterlies and the mid-latitude westerlies is known as the Polar Front, and this tends to follow the position of the upper-atmosphere waves already described (*p. 9*). It is at this frontal zone that most of the very active weather systems are generated. On a more limited scale, warm and cold fronts separate the various air masses within a low pressure system, and these will be described in detail later.

Anticyclones and depressions

Anticyclones, with their generally settled weather, are none the less active in that they extend their influence in the form of ridges, which may then decay, as well as shifting their centres. Their movements may be erratic and they frequently move from east to west rather than in the more normal west to east direction. Overall however they tend to follow paths eastward and towards lower latitudes (equatorward). Such highs may be quite short-lived – 4 or 5 days – passing rapidly across any region, but others, on rarer occasions, may become very persistent and cover an area more than 1500 km (900 miles) across, and last for a period of weeks. Such large systems are known as **blocking highs** as travelling depressions are diverted onto paths around the northern or southern flanks of the stable air system.

The variations in the area under anticyclonic influence as well as its general movement mean that wind directions may slowly change, and different air masses be brought into a particular locality. However any changes will tend to be minor ones which do not disturb the overall settled pattern. Nevertheless this settled pattern does not mean perfectly clear, cloudless skies all the time (although these are often found, especially in summer, and in cold continental interiors in winter). However, in winter particularly, the subsiding air forms an inversion (*p. 32*) at moderate height and a heavy layer of strato-cumulus cloud (*p. 38*) may build up. Pollution in the form of smoke or smog (*p. 85*) will also contribute to a general anticyclonic gloom.

In contrast to the quiescent conditions in anticyclones, depressions – more correctly known as **extra-tropical cyclones**, distinguishing them from tropical cyclones such as hurricanes – bring changeable conditions, and are responsible for most of the 'weather' in the middle latitudes, so that the results of their influence form a large part of this book. Because of their importance, it is helpful to consider the formation and structure of these low-pressure systems in some detail.

Depressions are generated at the Polar Front, and to a lesser degree at the Arctic (and Antarctic) Front. Depressions generally travel eastward and poleward, except when steered around highs

Fig. 12 The formation and development of a depression. Three-dimensional representations are shown on the left, and corresponding isobars and fronts on the right. At first the winds flow parallel to the isobars **a**, and for a while continue to do so as a slight wave begins to develop **b**. Air starts to flow across the isobars and a distinct low pressure area forms **c**. The cold front is beginning to overtake the warm front as the air in the warm sector rises above the cold **d**. The cold front has reached the warm front and given rise to an occlusion, and a pool of warm air has been lifted away from the surface **e**. The system then begins to disperse.

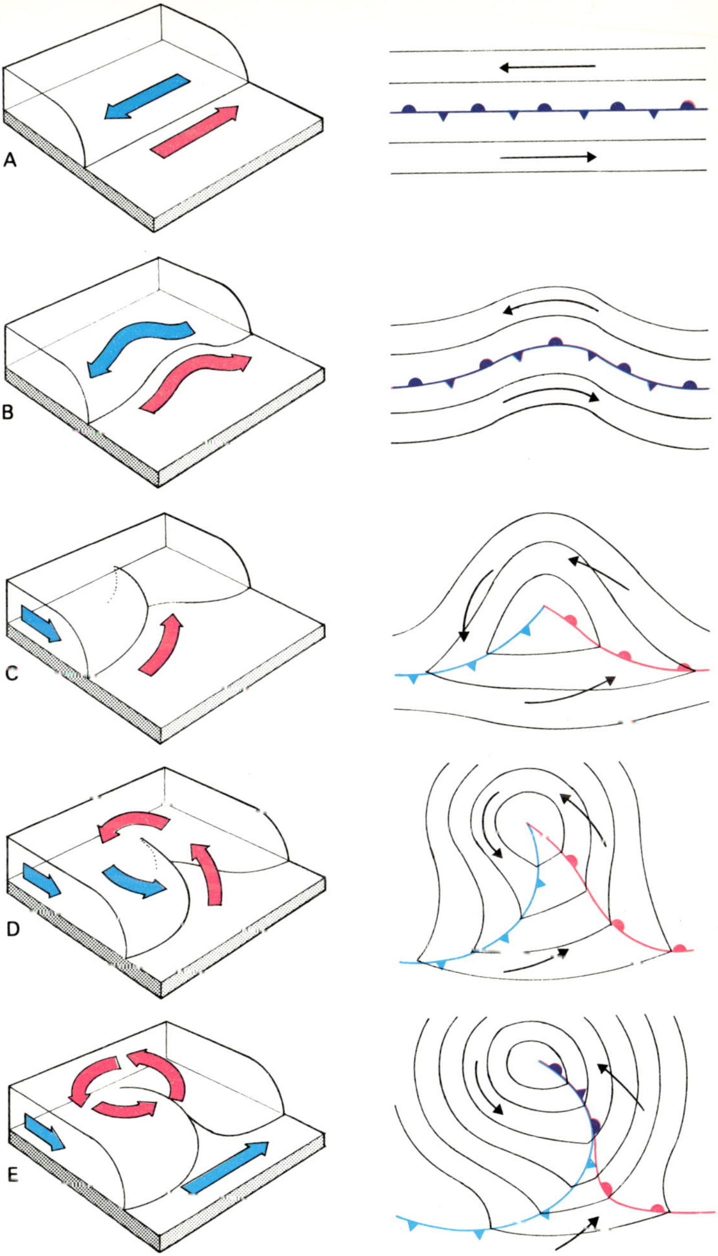
A
B
C
D
E

as just described. A low-pressure system forms when a wave develops at the boundary between cold and warm air flowing in opposite directions as shown in *Fig. 12*. A distinct structure appears, with warm and cold sectors preceded by warm and cold fronts respectively. The low pressure region develops with its cyclonic circulation and the whole depression tends to move in accordance with the winds of the warm sector. The warm air overruns the cold at the warm front and typically ascends with a slope of 1:150. At the cold front the cold air has a more vigorous undercutting action, and the slope is more usually about 1:75 – twice as steep, thus wind speeds are generally greater in the cold air which thus encroaches on the warm sector, so that the cold front soon overtakes part of the warm front and here the warm air is lifted away from the ground; a process known as **occlusion**. An occluded front therefore has two cold air masses (one colder than the other) in contact at the surface, and warm air aloft. The

Fig. 13 A family of depressions in various stages of development. There is a tendency for two mature systems partly to circulate around one another as shown on the right, where the second depression is overtaking the first on the warm-air side.

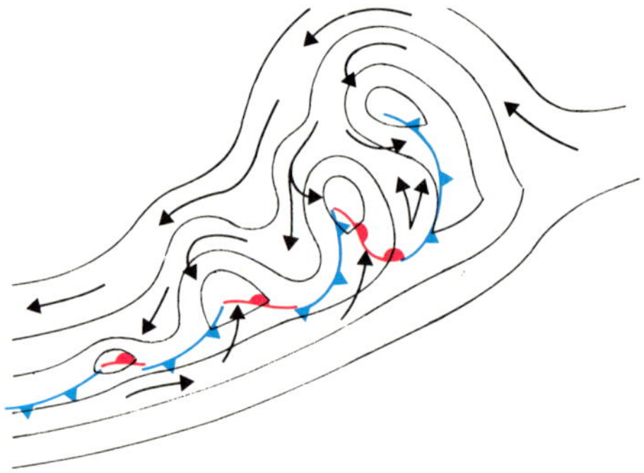

Dark stratocumulus clouds typical of some anticyclonic conditions. Banding of the clouds is evident both parallel to, and across the wind direction.

cold air encroachment continues until the whole front is occluded, when the system dissipates after a life of about five days. A whole family of depressions may develop, each on the trailing cold front from its predecessor, and further towards the equator. However generally each succeeding depression is weaker as the temperature contrasts decrease towards low latitudes, so the sequence breaks off, and a new family forms at much higher latitudes.

The 'classic' form of frontal system has been described, where the warm air rises along both frontal zones, which are technically known as an **ana warm front** and an **ana cold front** – the Greek word *ana* meaning 'going up'. However the warm air may be subsiding along part of any frontal zone (or at a particular level) which gives rise to **kata warm fronts** and **kata cold fronts** – from the Greek *kata*, 'going down'. The vertical motion at the various fronts, including the occluded types, give rise to extensive cloud systems described later.

Atmospheric layers and temperatures

The atmosphere is divided into a number of layers, mainly on the basis of the way in which temperature changes with height (*Fig. 14*). It is within the lowest layer, the **troposphere**, that there is enough water vapour for clouds to form. The altitude of the temperature minimum, the **tropopause**, which forms the upper boundary of the troposphere, is higher at the equator than the poles. Approximate average heights are 16-18 km (10 miles) over the equatorial regions, about 11-12 km (7 miles) at latitude 45°, and 8–9 km (5 miles) at the poles. Like everything else associated with the atmosphere, the altitude of the tropopause is variable and can show sudden breaks and changes of height. Where the tropopause plates overlap, the sharp gradient of temperature produces the fastest jet stream, which will in turn produce the most mobile disturbed weather.

It is within the troposphere that there is sufficient water vapour for cloud formation when conditions are appropriate. In the stratosphere above, and at still higher altitudes, the air is far too dry for the usual types of clouds to form. Two rare varieties can sometimes be seen from fairly high latitudes at sunrise or sunset. The **nacreous clouds** are one type, visible shortly after sunset, or before sunrise, and occurring at heights of about 24 km (15 miles). They appear to be a form of lee-wave cloud (*pp. 60-61*) and to be associated with particularly strong high-level winds and deep depressions.

The other variety is that of the **noctilucent clouds** which occur at

Even close to the equator the decline in temperature with height enables snow and ice to persist on high mountains. Kilimanjaro, shown here, is situated at latitude 3°S, and is 5889 m (19 321 ft) high.

This famous photograph of the eruption of Vesuvius in March 1944 illustrates how volcanoes may inject large quantities of dust into the troposphere or even the stratosphere where it may act as condensation nuclei.

altitudes of about 80 km (50 miles), around the upper boundary of the mesosphere. Although both types may give clues to the conditions at great altitudes, they are unimportant when studying surface weather.

As can be seen from *Fig. 14* the temperature falls more or less steadily with increasing height throughout the troposphere, so that at about latitude 45°, for example, with a ground level value of around 15°C (59°F), the temperature at the tropopause height of about 12 km (8 miles) is typically in the region of −55°C (−67°F). The rate of decrease, or **lapse rate**, averages about 0·6°C per 100 metres (6°C per km or about 17°F per mile). It is for this reason that it is frequently freezing cold on high mountain tops even in the warmest weather, and also why clouds at the highest altitudes are formed of ice crystals.

Fig. 14 The generalized variation in temperature with altitude is shown here. As explained in the text the exact height of the various layers varies considerably, as does the actual temperature profile.

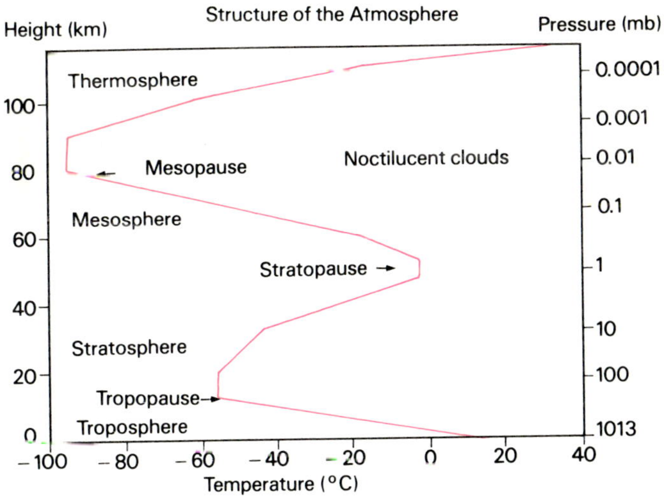

The formation of clouds

Although more technical methods of forecasting will be described later, the clouds and other phenomena visible in the sky give many valuable indications of the forthcoming weather.

Clouds act to harmonize the heat balance between incoming solar radiation and outgoing terrestrial radiation. Suppose the intensity of solar radiation changed. Radiative equilibrium would be restored by a change in the cloud cover. Increasing solar energy would be met by increasing cloud cover and rain. Decreasing solar energy would be met by decreasing cloud cover and rain.

Clouds form because the originally cloud-free air is cooled so much that the water vapour it contains condenses (at its **dewpoint**) on to dust or salt particles (**nuclei**) suspended in the air, forming liquid water droplets or if the temperature is low, freezing to produce ice particles. Atmospheric water droplets are frequently in a **supercooled** state, that is they remain liquid at temperatures well below 0°C (32°F). However below −40°C (−40°F) only ice crystals exist. Some fairly high clouds consist of supercooled droplets alone, but others contain all three forms of water, particularly cumulonimbus clouds with great vertical extents.

The most common methods of cooling are those where air comes

Fig. 15 *below* Under clear conditions (*left*) the ground surface can not only receive the maximum amount of solar radiation, but can also radiate heat away to space. When clouds are present (*right*), although less direct heat will be received, the clouds also act to absorb outgoing radiation, and re-radiate some of this back to the surface.

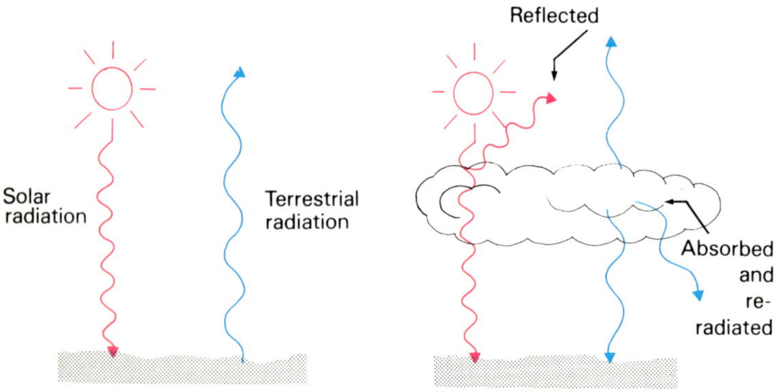

Fig. 16 *opposite* The three major processes of cloud formation are illustrated here. Convectional cloud is formed (*top*) when thermals lift warm, moist air above its condensation level.
Frontal cloud (*middle*) forms when two air masses at different temperatures are in contact, the colder undercutting the warmer. Finally, a mountain barrier may also uplift warm air (*bottom*) to give rise to orographic cloud.

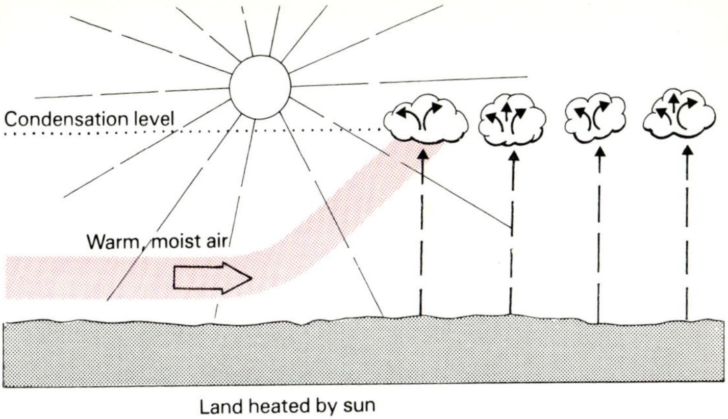
Condensation level
Warm, moist air
Land heated by sun

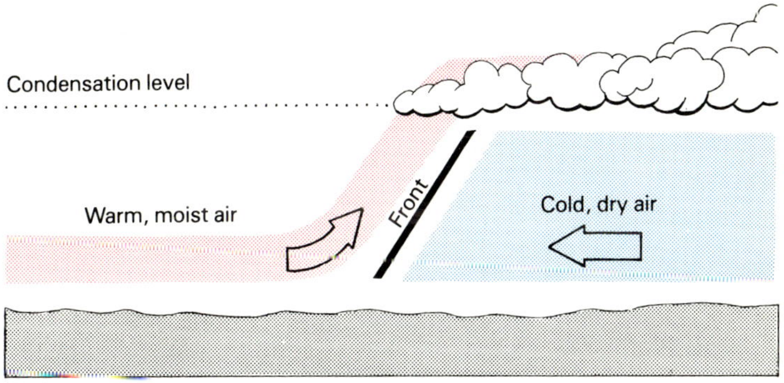
Condensation level
Warm, moist air
Front
Cold, dry air

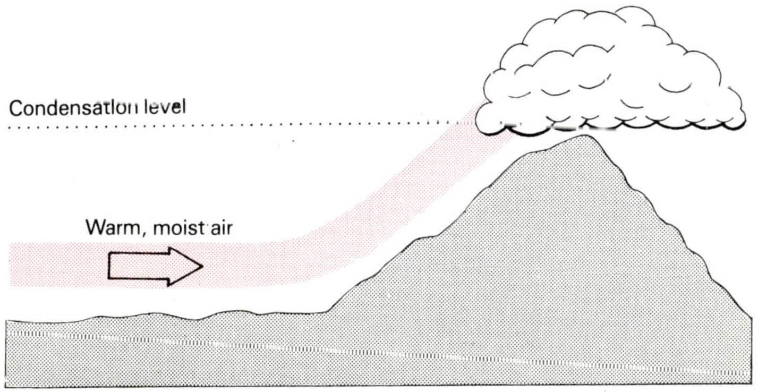
Condensation level
Warm, moist air

into contact with a cold surface below the dewpoint, most typically producing forms of fog (p. 74), or when air rises in the atmosphere. In the latter case this may be caused by convection, as for example, in cumulus and cumulonimbus clouds (*pp. 34-37*), by uplift over mountain barriers or more importantly, in low pressure systems, by a cold air mass moving beneath a warm, moist one.

The overall lapse rate of the troposphere is an average and is less than the amount by which a mass of dry air will cool by expansion when it ascends, or conversely, by which it will warm on descent. This, known technically as the **dry adiabatic lapse rate**, is almost exactly 1°C per 100 metres (10°C per km or 29°F per mile). The actual lapse rate over any vertical distance within the troposphere may be greater or less than this figure. If a certain mass of dry air is heated and begins to rise, it will continue to do so all the time the surrounding air has a greater lapse rate and thus remains the colder of the two. These conditions are said to be **unstable** as the vertical movement, once started, will continue. However if the actual lapse rate is less than the dry adiabatic, the mass of air will reach a position at which it becomes colder than the surroundings and will therefore begin to sink. Such conditions are described as **stable**.

For air saturated with water vapour the situation is similar, but the rate of temperature change, the **saturated adiabatic lapse rate** is less, varying between 0·4°C per 100 m (4°C per km or 11°F a mile) at high temperatures, and close to 0·9°C per 100 m (9°C per km or 27°F a mile) at −40°C (−40°F).

If a rising mass of air is not completely saturated, it will initially behave like dry air. However at some stage it will cool to its dewpoint and from then on act as saturated air. It is at this condensation level that clouds become visible.

If at some height the temperature of the atmosphere increases, rather than decreases, with height, an **inversion** is created. This is very stable and it acts to inhibit both rising and falling air movement.

The great variations in temperature and moisture content of air within the troposphere mean that clouds can form at many heights. The standard classification divides them into three broad classes (high, middle and low clouds) depending upon the height of the condensation level at their bases. Within these classes there are ten distinct types and these are shown in the table with typical mid-latitude base heights.

As can be seen from the table, although the distinct types are usually found within the three height classes shown, some variation does occur from time to time, so that they appear at somewhat greater or lesser altitudes. Layer, or stratiform clouds, are associated with stability and contrast with cumiliform clouds, such as cumulonimbus, which have great vertical extent and form under unstable conditions.

Cloud Classification

	Ht of Base (km)	Type (Genus)	Abbreviation
		Cirrus	Ci
High	5–13	Cirrocumulus	Cc
		Cirrostratus	Cs
	2–7	Altocumulus	Ac
Medium		Altostratus	As
	1–3	Nimbostratus	Ns
	½–2	Stratocumulus	Sc
Low	0–2	Stratus	St
	½–2	Cumulus	Cu
		Cumulonimbus	Cb

Fig. 1 / The heights of occurrence of various forms of cloud. It should be noted that the ranges of high and middle clouds overlap, and that cumulonimbus clouds can have great vertical extents.

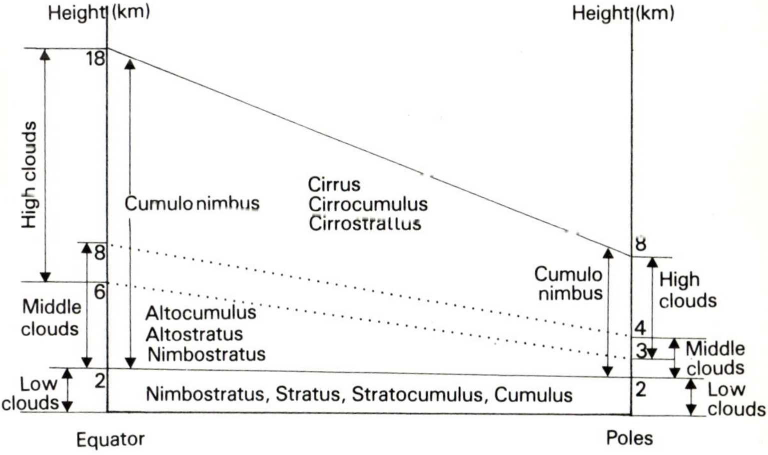

Cumulus and cumulonimbus

Of all the clouds the best-known form is **cumulus**, and on its own this type represents fair weather. With a little imagination these clouds can often be seen to resemble bizarre cauliflowers and this results from the way in which they are formed by convection in rising bubbles of hot air or **thermals** (*Fig. 18*). It is these thermals which are so sought after by glider pilots to give them lift in the lower layers of the atmosphere. The thermals are created as the earth warms up in the morning and as the temperature climbs small puffs of cumulus will merge and grow. Growth will be sustained so long as the air above the cumulus top is not warming and sinking (due to, say, rising pressure) and the rising air flow is itself moist. As the surface air temperature peaks in the early afternoon, cumulus reach their greatest depth soon afterwards. They then tend to die away in the late afternoon and evening, and dissolve as the air and earth cool down during the night. Well spaced out, separate cumulus clouds indicate a dry spell. The airflow is either too dry, or the surface air too cool, for any showers to develop.

When cumulus clouds absolutely litter the sky early on in the day, showers are a possibility. Shower clouds are, technically, 'precipitating convective clouds'; strictly speaking they are **cumulonimbus** clouds – the big brother of cumulus. Cumulonimbus can grow right up to the bottom of the stratosphere where the top must spread out under

Fig. 18 The general circulation of air in a thermal is shown (*left*). When the condensation level is reached cloud forms, but the thermal will continue to rise, the amount depending upon its strength and the ambient conditions.

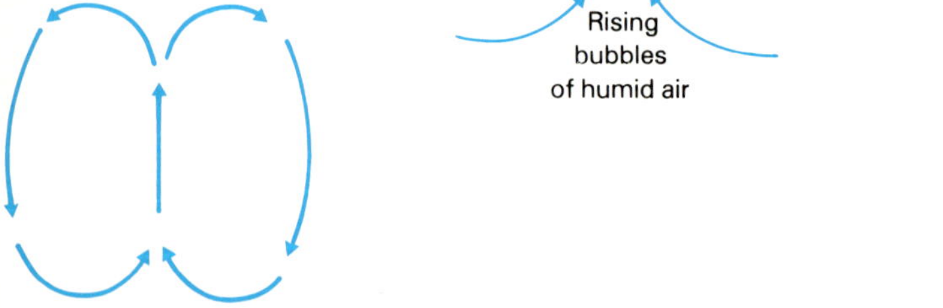

Small cumulus clouds forming at mid-morning. The distinct flat base seen on the more distant clouds marks the position of the condensation level.

Larger cumulus clouds which have built up by mid-day or early afternoon. Although the bases have darkened no rain is falling at this stage.

Opposite top A large bank of cumulonimbus clouds with dark bases. The cloud on the left has a fibrous top indicating that it is composed of ice crystals. The tower on the right is still actively growing.

Opposite below Cumulus and cumulonimbus building up over the East African Rift Valley. Active spreading is taking place in the layer just beneath the top of the large cumulonimbus cloud.

Fig. 19 Cumulonimbus clouds form from a series of cells. Here an old cell **a** has reached the tropopause and expanded into an anvil shape. A second **b** is just reaching that stage, while the next is forming to one side **c**.

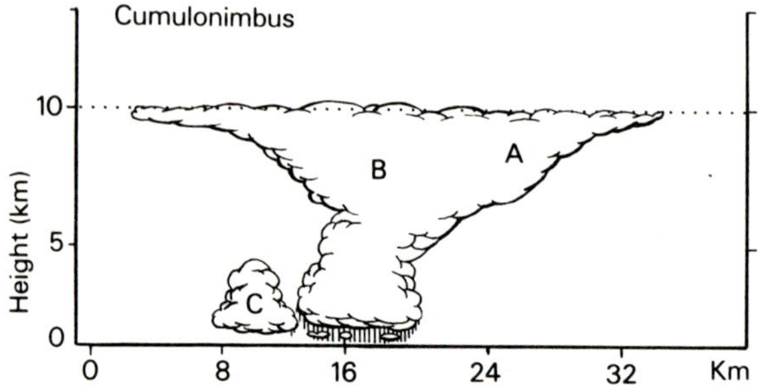

the warmer air above and so takes on the shape of an anvil, which may be visible at great distances. The tops of cumulonimbus clouds are frequently fibrous, being composed of ice crystals, and resemble cirrus (*p.44*). Spreading under a stable layer can also give rise to cloud greatly resembling altostratus (*p.42*) or nimbostratus (*p.48*) and which, especially if the wind increases aloft, can stream far away from the parent convection cells. When the convection cells feeding warm moist air into the cumulonimbus become inactive, and the system decays, it frequently happens that wispy ice clouds remain to outline the shape of the anvil, together with some patches of altocumulus or cumulus clouds scattered in the sky.

Even when they do not form anvils, cumulonimbus may form mountainous masses of cloud, and in either form sighting of cumulonimbus means that the air is moist for a considerable depth in the atmosphere, and that the surface air is a good deal warmer than the air immediately above. The great depth of cloud provides an ideal environment for water droplets to grow by condensation and coalescence, and for ice crystals to grow by stealing the water vapour evaporated from the millions of supercooled water droplets. Eventually the droplets grow so big and numerous they fall to the cloud's base, blocking the sunshine and darkening the bottom, and they then begin to fall to earth as rain. Black bottoms and anvil tops are sure signs of showers to come. The development of showers and thunderstorms and their prediction is more fully described on *pp. 98–105.*

Stratocumulus

The layer cloud closes related to cumulus is **stratocumulus** and it is indeed most frequently formed by the spreading out of cumulus cloud, and it may be regarded as a low form of anvil cloud. Its typical form is a flattened cumulus that looks something like a pancake. Usually there are gaps of blue sky visible between the individual cloudlets, but sometimes stratocumulus fills the whole sky as a formless, fairly bright blanket. Very frequently the clouds are arranged in bands which lie across the wind, thus indicating its direction at that height. However,

An extensive layer of stratocumulus with the remnants of a second lower layer. The banded structure can be distinctly seen, with the rolls lying across the wind which came from lower right.

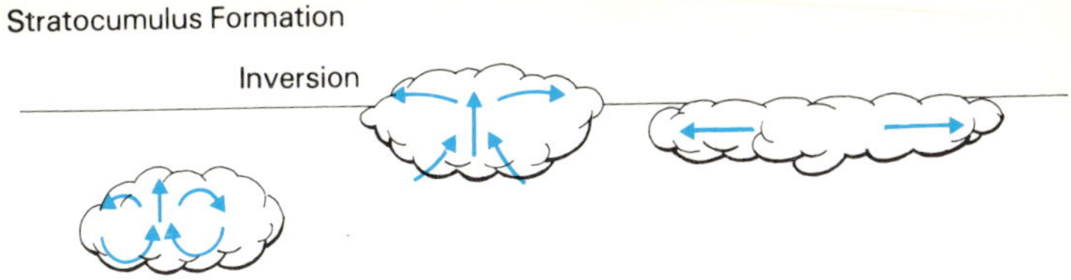

Fig. 20 Thermals rising through the condensation level (*left*) continue until they reach the inversion, which they tend to overshoot (*centre*), before flattening out sideways (*right*) when the stratocumulus base may be considerably above the dewpoint level.

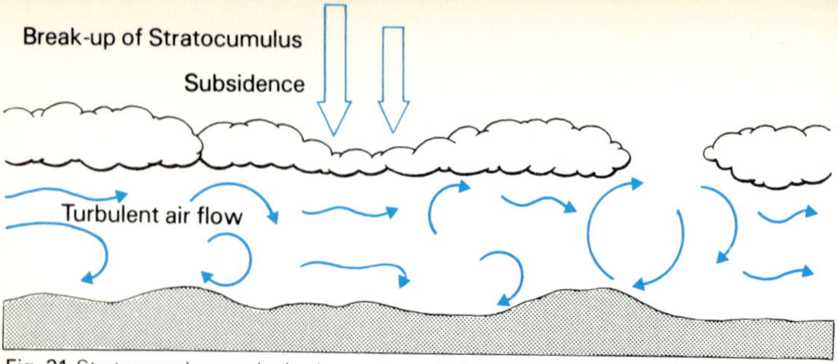

Fig. 21 Stratocumulus can be broken up by turbulent mixing of the lowest air layer, while thermals may also assist. Similarly a general subsidence and increase in pressure can dissipate the cloud layer.

stratocumulus only forms when the winds are light or moderate.

Stratocumulus is very common in some regions of the world, over Western Europe for example, where it very frequently exists in considerable amounts in the warm sector of low pressure systems, and especially in those systems where the warm air is subsiding. Sometimes very thick, dark layers of stratocumulus are formed, but despite their appearance, these do not give rise to very great quantities of precipitation.

This cloud type on its own signifies a dry spell where the temperature variation between night and day is only a few degrees. The weak thermals that create stratocumulus lose their buoyancy when they meet upper air that is just as warm. The saturated thermals are forced to spread out under the inversion, forming sheets of cloud which tend to have great stability. The base of the final cloud layer may be well above the condensation level after the thermals have spread out, so that the base appears quite sharp and distinct. Very strong thermals may sometimes penetrate through the inversion to give rise to cumulus clouds above the stratocumulus sheet. These may sometimes be difficult to recognise from the ground, being hidden, but they frequently give themselves away by the denser shadows at their bases.

Unlike cumulus or cumulonimbus, which have little effect on the general increase in temperature during the day, a blanket covering of stratocumulus means that the maximum daytime temperature is reduced to about three-quarters of its potential and so a persistent sheet will inhibit the formation of thermals and cumulus clouds the next day. At night stratocumulus acts like a blanket, trapping the Earth's heat and so maintaining the temperature.

One of the most difficult tasks of weather forecasting is to predict when, if at all that day, stratocumulus will break up and allow either the Sun's heat in or the Earth's heat out. Pointers to look for are the brightness of the day, the trend of the atmospheric pressure, and the wind. The brighter the day the thinner the cloud, and the greater the chance of breaks appearing. The larger the rise shown by the barometer, the stronger the subsidence eating away at the top of the cloud blanket, and the greater the chance of breaks appearing.

Cumulus cloud spreading out into stratocumulus. The condensation and inversion levels are distinct, as is the base of the actual stratocumulus layer.

However, strong pressure rises alone are usually no match for the stability of a very dull day full of stratocumulus and the main destructive process is the wind. Apart from the straightforward case when the wind simply blows the cloud bank away somewhere else, a turbulent surface wind is good news when the sky is overcast with stratocumulus. The stirring of the lowest layers of air eats away at the bottom of the cloud sheet and, given time, will eventually make blue holes in the cloud. But if this process takes until the afternoon to make any impression on the cloud sheet, the most likely result is a temporary increase in temperature, causing stronger convection; these new thermals simply grow up as cumulus and refill the holes that the earlier turbulence created.

If, on the other hand, the first breaks in the cloud sheet occur during the late afternoon or early evening when convection has died down, then usually the cloud will continue to break up to give a fine evening. This is because the daytime turbulent mixing of the air transported warmer air up and cooler air down, allowing the warmer air to evaporate the bottom of the stratocumulus sheet slowly without causing additional strong convection from the ground. In short, the dispersal of stratocumulus is a finely balanced problem.

Altostratus and altocumulus

The formation of middle and high clouds is usually brought about where horizontal convergence causes extensive layers to rise slowly, for example in low pressure systems where the warm air mass is rising at warm or cold fronts. The most immediate result of such uplift is the formation of the two layer clouds, altostratus and cirrostratus, both of which occur under stable conditions. **Altostratus** normally forms as a fairly uniform grey sheet of cloud, which consists of water droplets (frequently supercooled) or snowflakes. When it is thin the position of the Sun can still be seen, and if conditions are right it may be surrounded by a corona (*p. 54*). Very much thicker altostratus can completely obscure the Sun, and although usually little rain or snow reaches the ground from this form of cloud, altostratus frequently grades into the variety of cloud known as nimbostratus (*p. 48*) which can produce large quantities of precipitation.

Altocumulus clouds usually occur in distinct layers rather than individually, and take the form of white and grey cloudlets, which are fairly regularly arranged and appear distinctly rounded and shaded. They are often seen as more or less parallel rolls with clear lanes in between, and they generally lie across the wind at that height. As seen from the ground the sheet of cloud moves slowly as a whole. With greater winds at height the motion of the individual cloudlets becomes appreciable and the regular rolls break up and the cloudlets become arranged in bands downwind.

A typical sheet of almost featureless altostratus is here seen clearing after a period of rain. The layer is breaking up and thinning as the rainbelt passes away towards the right.

The regular patterns of altocumulus clouds may create very beautiful skies. They frequently develop from an even sheet when shallow convection sets in, overturning the air and breaking up the layer.

Thundery showers are often preceded by clouds known as **altocumulus castellanus**. Not as high as ice clouds and not as fat as normal cumulus, castellanus clouds are formed not by thermals but spontaneously by the heat released when condensation occurs (just the opposite to the cooling effect of evaporation). They look taller than normal cumulus and they are a sign that the medium level air is both moist and unstable, requiring only high surface air temperatures to spark off thundery cumulonimbus outbreaks in the vicinity.

This altocumulus castellanus shows towers growing upwards from the main layer, indicating instability and moist air at that height.

Cirrostratus, cirrocumulus and cirrus

Cirrostratus is a thin whitish sheet of cloud consisting of ice crystals which gives a milky appearance to the sky. It does not greatly obscure the light of the Sun and Moon, but these are frequently accompanied by halo phenomena (*p. 54*). The thickening of cirrostratus to altostratus is a good indication of an approaching front with its accompanying rain and bad weather.

Just as stratocumulus may be broken up by turbulence and convection, so may altostratus and cirrostratus cool sufficiently by radiation for the lapse rate within the layer to exceed the saturated adiabatic rate. When this happens shallow convection begins and holes appear in the cloud sheet, which may eventually disappear.

Cirrocumulus clouds greatly resemble altocumulus in their structure and arrangement, but the individual white cloudlets appear much smaller. The sheets and patches of cloud show fine ripples and grains often arranged in bands, like altocumulus, by the action of upper level winds. Their height is so great however that their motion is not easily seen from the ground.

Cirrostratus of a more localized type than the usual veil over the sky. Wave motion caused by hills at a distance to the bottom right probably accounts for both the position and structure of the main cloud.

A typical layer of cirrocumulus. The fine, regular structure in the sheet is often seen, but unlike altocumulus there is no distinct shading to the clouds.

Cirrus clouds are characteristically feathery or fibrous in appearance and consist of **fall-streaks** of ice particles. They very frequently appear by the direct formation of ice crystals from apparently clear air or by the freezing of a previously existing altocumulus or cirrocumulus cloud. The trail of ice particles may appear to come from a distinct denser head or may be stretched into long streamers several kilometres or miles long. When cirrus is arranged in long dense bands it shows distinctly the position of the high-speed jet streams, while gradually increasing cirrus cover is a sure sign of the approach of a warm front.

Characteristic cirrus clouds. Streamers of ice particles are falling from the generating heads into slower moving layers of air. Ahead of a depression, the trails point towards the warm air mass.

Clouds and precipitation

Precipitation is the term used by meteorologists for any kind of solid or liquid water-based deposit from the atmosphere. In other words it includes rain, snow, drizzle, hail, fog, frost and dew. For the present we shall confine ourselves to examining those types of precipitation that fall to the ground from clouds, i.e. drizzle, rain and snow (hail is a rather special case and is described separately on *p.101*).

It is common knowledge that not all clouds produce surface-level precipitation. In the case of high-level clouds, precipitation may be produced but it has so far to travel before reaching the ground that it has time to evaporate on the way. On the other hand, shallow, low-level layered clouds produce little precipitation in the first place, because of their very limited depth. The type of cloud known as **stratus**, which frequently covers the sky with a featureless grey sheet, but through which no optical phenomena are seen, will only produce drizzle or very fine snow if it produces anything at all.

Relatively shallow convective clouds produce no precipitation because they do not reach up to regions with cold enough temperatures. To understand this it is necessary to know the two processes by which precipitation comes about. The first process, **coalescence**, requires the tiny condensed water droplets and supercooled water droplets within the cloud to merge together and form very much bigger raindrops. In the case of layered cloud, the droplets can spend so much time within it that they eventually coalesce and grow bigger through random motion; in a convective cloud the droplets are transported so fast by the turbulent air currents inside the cloud that they collide and grow bigger, again through coalescence. In addition, large drops fall faster than small ones and so they tend to sweep up the small droplets below them.

The second process – which takes place in the higher (colder) parts of the cloud – involves ice crystals growing as they pass through regions of supercooled water droplets. The supercooled droplets are constantly evaporating and condensing: a sign of this is the way in which cumulus cloud is continuously changing shape. The water vapour condenses more readily on to ice than on to water and the ice crystals therefore grow into **snowflakes** that fall through the cloud, sometimes melting into large raindrops.

The speed at which precipitation leaves the cloud determines how long it spends in the relatively dry air intervening between it and the ground. Layered clouds are usually gentle affairs where precipitation creeps out gradually, saturating the intervening air, where raindrops often reduce to **drizzle** (when the drops are only just large enough to fall), and snowflakes to a mixture of snow and rain. (Outside the United States of America this mixture of melting snow and rain is termed **sleet**, but American practice uses this word to describe the ice pellets formed when rain falls into a colder layer of air and freezes.)

Altocumulus with some fallstreaks, or virga. The clouds are moving from the left and precipitation, which is falling into a slower layer of air, is evaporating before it reaches the ground.

Stratus cloud shrouding the top of Mount Pilatus at the western end of the Lake of Lucerne in Switzerland. A heavy rainstorm has just passed over and is clearing away.

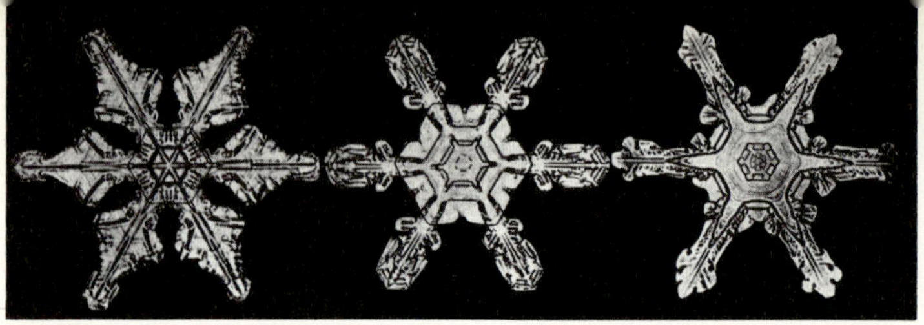

Snowflakes present an endless variety in the forms which have been produced during their growth. Apart from the type shown here, hexagonal plates and columns are also quite common forms of ice crystals.

Convective clouds such as cumulonimbus literally throw out the precipitation, but if the intervening air is warm enough the snow will melt all the same and reach the ground as rain or a mixture of rain and snow. Rain (or drizzle) which freezes on coming into contact with the ground or any other object at a low temperature is known as **freezing rain**.

Apart from cumulonimbus the main rain- (or snow-) bearing clouds are the layer clouds altostratus (*p. 42*) and nimbostratus. These may be found either together or separately in low pressure systems, where the altostratus very frequently thickens and changes to the lower nimbostratus. However it is only when altostratus is thick that large quantities of rain are produced, so that when the position of the Sun is still vaguely discernible, rain is probably not actually falling, though it is doubtless imminent.

Nimbostratus, on the other hand, is very low, dark grey cloud, through which the Sun is never visible. From it rain (or snow) falls

The clearance of frontal rain clouds over the Badlands National Monument in South Dakota. The active front passing away to the left is followed by skies with both ice crystal cirrus and water droplet cumulus clouds.

more or less continuously. It is frequently accompanied by ragged shreds of cloud, partly or completely detached, beneath the main cloud-base. In contrast to shower clouds which may give very heavy rainfall, but which are of fairly short duration, nimbostratus often gives many hours of practically continuous, heavy rain or snow.

In the polar regions snowfall is light as the cold air contains little water vapour, and for the same reason, amounts in the interior of large continental land masses are only moderate. However those regions where warm maritime air streams encounter cold continental air masses can be expected to have large (if less frequent) falls.

The classic snow situation in winter is when a cold air stream brings dull grey overcast skies and pushes the surface-level air temperature below 3°C (37°F), allowing snow to reach the ground and accumulate. Above the cold wedge of layered cloud, moist tropical air rides upwards to the coldest regions of the troposphere, where snowflakes form in their untold millions. The snow then persists until the cold air flow ends and the higher temperature turns the snow to rain. At temperatures close to freezing, small crystals tend to freeze together and form large flakes of wet snow which can cause great inconvenience when the surface temperature is about 0°C (32°F), as it will commonly melt in sunshine during the day, or under pressure such as that from road traffic, but will refreeze into ice at night. In contrast, at lower temperatures dry snow will form and as this consists of very small crystals which stay separate, it can therefore be easily moved by snow blowing machines provided temperatures stay low.

Precipitation from stratus cloud is usually light but persistent. In this example it can be seen that much of the rain is evaporating before it reaches the ground.

Sky and cloud colour

Normally the sky is blue because the incoming sunlight is strongly scattered by the tiny molecules of the air, and this affects the smallest wavelengths (the blue and violet) by far the most. So the indirect light from all around the sky is rich in blue (in addition to which the human eye is more sensitive to blue than violet). Normally the Sun is yellow because, with all the blue and violet scattered out, a predominance of yellow (and sometimes red) is left to reach the eye directly.

Cloudless skies can reveal quite as much about the weather as those which are full of clouds. When there are no clouds, it is the shade of blue which is important. A clear deep blue indicates that the air is too dry for clouds to form, although almost invariably there are small amounts of cloud on the horizon. In fact there always seems to be more cloud on the horizon than overhead because of perspective.

Hazy, light blue, or silvery-blue skies indicate that the air is too stable to allow clouds to form. The sinking air associated with the high pressure puts a 'lid' on the possible depth of convection, which, being lower than the surface air's condensation level, keeps the skies cloud-free. The pale colour of the sky is brought about by dust and other pollutants being trapped below the upper limit of convection in such high concentrations; this **haze** has such a range of particle sizes as to scatter the incoming visible light components at random, so that the colours overlap and the indirect light from all around the sky appears much brighter than usual.

Similarly, clouds are normally white because their collections of

Fig. 22 The size of the molecules in the air is such that they strongly scatter blue light in all directions so that this is all that is seen by an observer at A. Red light is usually unaffected so that at sunrise and sunset, for example, it may be all that an observer at B will see.

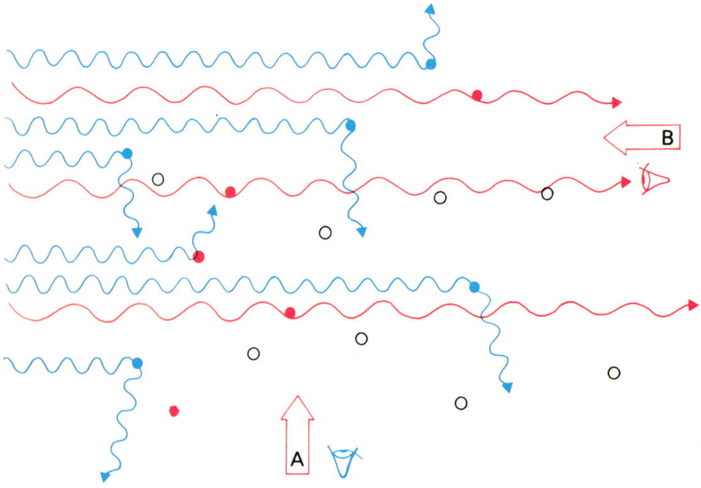

Most scattering of blue light is caused by the molecules of the air itself. In this Alpine view larger particles are also scattering light to give haziness increasing with distance.

A cloudless sunset showing some of the gradation of colours. The low sun is seen through a haze layer, built up under an inversion during the day.

droplets and crystals are of all sorts and sizes, causing the colours around each one to overlap destructively so that all colour disappears. Although it is obvious that clouds which are in shadow will appear dark, the effect is accentuated by contrast so that some clouds seem almost black when seen against those which are brilliantly lit. Quite apart from the yellow and red colours at sunset and sunrise, unusual circumstances can sometimes produce brilliant shades in clouds, so that completely blue clouds, for example, can occasionally be seen.

The colour of the sky at sunrise and sunset is sometimes a magnificent spectacle, and can provide some clues about the weather to come. As the Sun sinks in the sky its direct light travels further and further through the atmosphere to reach the observer. The white light therefore encounters many more air molecules than it did before and the scattering greatly increases, so that the light reaching the observer consists of a relatively high proportion of yellow and orange. At sunset the light has the maximum distance to travel through the lowest, and typically the dustiest, layers of the atmosphere. The dust increases the amount of scattering so that the Sun appears to turn from orange to red and illuminates the western horizon with a red glow. The sky further

away, lit by indirect light, is still rich in blue so that somewhere between the blue area and the red horizon the sky takes on a purple coloration. In other parts of the sky the particles may be just the right size to allow the green and yellow parts of the spectrum to stand out. As the critical factor in determining the predominant sky colour is the size of the scattering particles, only on one or two rare occasions have dust particles of just the right size scattered the red component of moonlight more than the blue – literally once in a blue moon.

A bright 'red sky at night' with the undersides of high cloud being lit by the sinking sun, indicates clear sky and dry air to the west. This is typical of the clearing weather following the passage of a cold front, and as long as winds are from the west, as is usual, it is most unlikely for a further front to follow before the next day and bring bad weather. However, a dull red sunset sky which is quickly blotted out by cloud indicates, again with a westerly wind, that rain can be expected soon.

A 'red sky in the morning' (rather than just a red sun disc) means that high-level ice clouds are already invading the sky and being illuminated by the rising sun. If the wind at cloud level is from the west, a warm front with its attendant rain clouds is likely to follow.

The red tint beneath these moderately high clouds illuminated by the rising sun shows that the air to the east is clear and fairly free from moisture.

Sunset over Hallendale Beach, Florida. The orange, rather than red, coloration suggests that a deterioration in the weather can be expected within a few hours.

Optical phenomena

Most of the optical phenomena that occur in the atmosphere around us – the unusual colours and luminescences – are fascinating and often beautiful, but not necessarily significant as far as weather forecasting is concerned. The following are a few that may be useful.

A **halo** is a bright circle appearing round the Sun, or occasionally the Moon. The angle between the Sun and the halo as seen by the observer is 22°, and the area of sky between the two is darker than that outside the halo. The halo is white except for a faint red tinge on the inside and violet on the outside; when a halo appears around the moon the colours are too weak to be seen. Often the halo is brightest at the top and bottom. The occurrence of a halo indicates the presence of ice crystals, as it is caused by hexagonal ice crystals which refract visible light through a minimum deviation angle of 22°. Normally this means that high-level cirriform ice clouds are invading the sky – and these are the forerunners of warm-front rain. In winter on a foggy day it can also mean there are ice crystals in the fog, so that especially at night hazardous freezing fog can be expected. There are many other phenomena related to haloes, but the most common are brilliantly coloured bright spots level with the Sun, known as **mock suns**; these are produced in the same way as haloes and so confirm the presence of high-altitude ice clouds.

A coloured ring around the Sun, smaller than the halo, is known as a **corona** and indicates the presence of shallow cloud composed of water droplets or ice crystals which are much smaller than usual. On a foggy day the appearance of a corona indicates that the fog is thinning and should soon break.

Above A fairly brightly coloured mock sun, frequently seen (as here) with a white horizontal tail extending away from the direction of the Sun to the left.

Left A corona in altocumulus over Kenya, the Sun's light being reduced by an imminent total solar eclipse. The exhaust trail of a scientific rocket just launched can also be seen.

Far left The 22° halo around the Sun, seen through ice-crystal cirrostratus. On the right part of a tangential arc, one of the many other halo forms can also be seen.

Fig. 23 The glory always forms around the point marked by the shadow of the individual observer's head (the anti-solar point), and is frequently seen from airplanes when a fogbow may surround the glory.

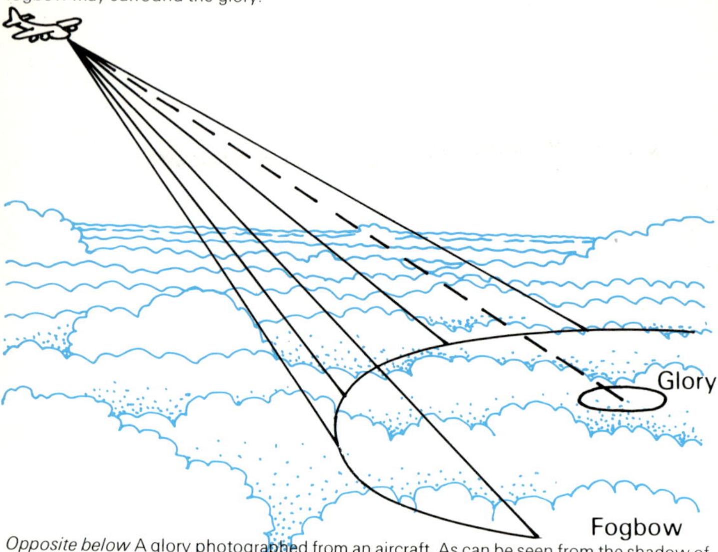

Opposite below A glory photographed from an aircraft. As can be seen from the shadow of the plane, glories always form directly opposite the Sun in the sky.

Brilliant patches of green and pink on clouds within about 40° of the Sun are known as **iridescence**. When they are seen in medium-level clouds such as altocumulus they are probably most striking and then indicate that the air is rising slowly on a large scale and are thus a sign of medium-level instability. Iridescence is actually very common in many clouds, but is very frequently overlooked.

While we ordinary mortals do not have haloes round our heads, we may on a bright clear morning see a bright white patch around the shadow of the head. This, the **heiligenschein**, occurs when a surface such as grass is covered with dew, the dewdrops being just the right size and position to focus on to the leaves an image of the Sun, which we then see magnified through the dewdrops. Although it is an interesting phenomenon all it really reveals is that the ground is dew-covered, which is obvious anyway.

Very rarely, colours can be detected on the dew in similar circumstances. However, this phenomenon, known as a **glory**, is more commonly seen around the head of the observer's shadow cast onto a fog bank. It indicates not only that the fog is thick (again, this would be obvious) but that the droplets are very small. A glory is frequently seen from an aircraft flying over clouds composed of water droplets and is very often accompanied by a white cloud- or fogbow (*p. 58*) although this is sometimes too faint to be readily perceived, unless special care is taken to look for it about 40° away from the glory.

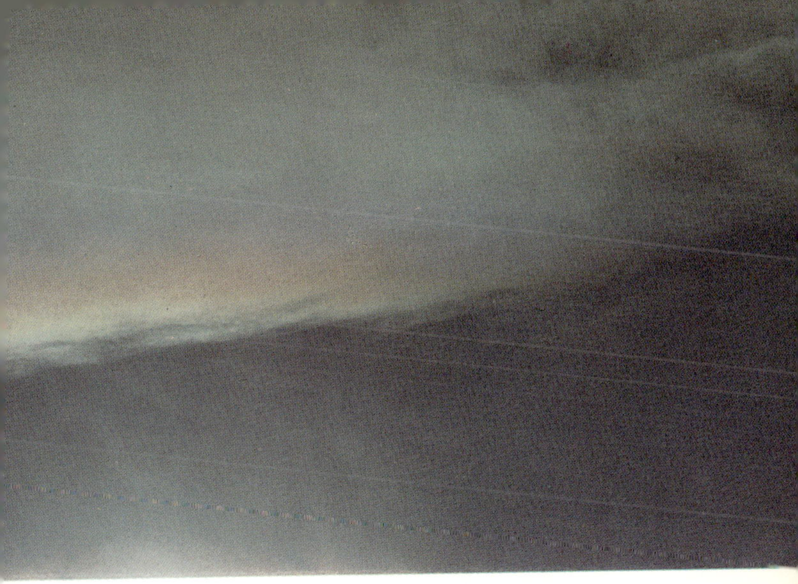

Above Iridescence along the edge of a sheet of altostratus cloud which was breaking up into altocumulus just outside this picture area. The Sun was several degrees below the bottom of the picture.

Rainbows and fogbows

One of the commonest optical phenomena, and the one most celebrated in verse, is the **rainbow**. Everyone knows what a rainbow looks like – a coloured arc of a circle produced on a screen of raindrops, most brilliant when produced by the Sun, much weaker when it appears by moonlight. The colours of the rainbow are caused by the refraction reflection and refraction again of sunlight within tiny spheres of water. The distribution of sizes in the raindrops determines how much, if any, of each colour is present. It is relatively rare for all the conventionally named colours – violet on the inside, indigo, blue, green, yellow, orange, and red on the outside – to be distinguishable and in any case they grade into one another. When viewed from a very high vantage point the rainbow may be a complete circle but the size may be regarded as constant at 40° (inside) to 42° (outside) radius. (This applies to the bright **primary** bow.)

When the outside limiting colour of the rainbow is noticeably red rather than orange, then amounts of rainfall are likely to be large; conversely when the red and orange bands are weak the amount of rainfall will be much smaller. This is because the largest drops refract red light (the colour with the longest wavelength), in a more pronounced manner, producing a bright band of red. The smallest drops are those found in fog. Here the colours created by the droplets all overlap and become lost, so that a **fogbow** is almost completely white with just a tinge of colour on the inner and outer edges.

A further bow with an average radius of approximately 51° is quite frequently seen. This **secondary** bow, which never appears on its own, has a reversed sequence of colours with red inside and violet outside. The colours are explained by a double reflection of sunlight within drops of rain and are therefore considerably fainter than the primary bow. The space between the bows is noticeably darker than the surrounding sky.

Opposite A bright, shallow rainbow photographed in the Canary Islands. The lower the top of the bow, the higher the Sun's altitude must be – in this case it is about 30° above the horizon.

Fig. 24 The primary rainbow is formed with a single reflection inside the water droplets, while two reflections produce the secondary bow. In both cases the various colours are separated by the effects of refraction.

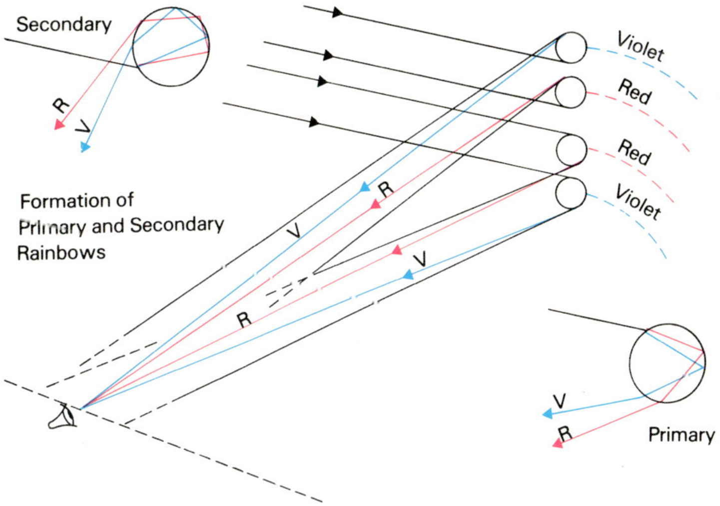

Mountains, hills and air waves

It is important to interpret general weather patterns in the context of where you live. Although large weather systems tend to affect areas of thousand of square kilometres or miles, there are certainly local variations caused by the lie of the land, sometimes creating many unusual microclimates in a comparatively small area. So as well as working out the broad weather picture it is necessary to be aware of one's situation in relation to the surrounding country, particularly in regard to mountains or hills.

The effect of the Earth's surface upon the wind is to retard it by frictional drag. Obviously the greater the surface roughness in any form of obstacle (whether it be buildings, trees or hills) the more the stream of air is affected, with the least friction being over the sea. The turbulence generated at the surface thoroughly mixes the air in contact with the ground, so that the lowest layer is very uniform in its temperature and moisture content. Away from the surface, in the free air above, the wind blows more freely. However on meeting hills or mountains across its path, air is forced to rise and accelerate, spilling over the crests of the hills and rushing down the leeward slopes. With a slow air flow the effects may be little more than increased turbulence, but with faster air streams even a fairly small range of hills will produce regular waves that remain fixed with respect to the obstacle. These lee-waves can exist for very considerable distances behind the hills, dying out downwind. Obviously winds blowing more or less at right angles to a range are the most effective in producing wave motion, and glider pilots, who frequently use the local lift generated on hill slopes facing the wind, are even more concerned to watch for the wind-shifts or changes in strength that may cause (or destroy) the wave motion

Fig. 25 The effects of friction are least over the sea (*left*), moderate over normal land surfaces, but quite considerable over built-up areas, and even greater in mountainous country, where strong turbulence can be produced.

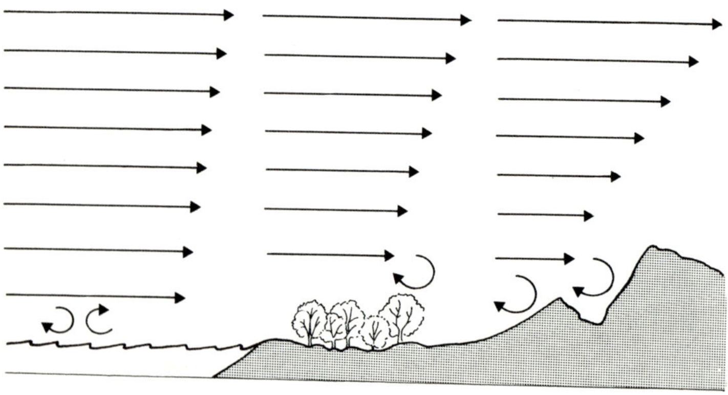

Wave clouds (altocumulus lenticularis) formed by the wind flow over the Chiltern Hills in England. Such clouds remain stationary in the sky all the time the wind speed stays constant.

which provides their much-sought lift at higher altitudes.

At the crests of the topographically induced waves the air will often rise and fall through its condensation level, so that humps of wave cloud form, frequently at more than one level, depending upon the conditions in the various layers of air. Even when no wave clouds form to the lee of a ridge of hills, the invisible turbulence is still there, and this turbulence is responsible for breaking up sheets of stratocumulus locally, although elsewhere over flat country the same type of cloud may remain unbroken.

Fig. 26 The formation of lee waves and wave clouds. The train of waves may extend far downwind and when the terrain is complex waves from one ridge may reinforce or cancel those from another.

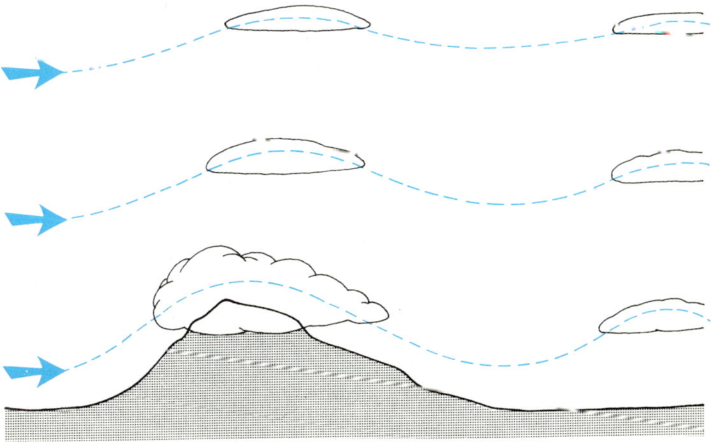

Hills, clouds and rain

Cloud will often crown the top of a hill while the rest of the sky is almost cloudfree. Surface air on the windward side is forced to climb the hill. As it does so it loses its capacity to hold its vapour and leaves its moisture as hilltop cloud with the cloud base lower on the windward side.

This **orographic** cloud (*Fig. 27*) will be stratiform under stable conditions, in which case it will hang over the peaks themselves and persist for a moderate distance to leeward, dispersing as the air warms with its descent on the other side. In unstable air cumulus-type clouds will form, again usually only persisting over the high ground. It will obviously depend upon the exact conditions at the time whether the orographic clouds will shroud the summits in stratus or other clouds, or whether clear air will remain below the cloudbase. Anyone likely to be out in hills or mountains would be well advised to be cautious when a warm, very humid air flow is expected to rise over the hills, as quite apart from the likelihood of poor visibility, falls of rain or snow in greater or lesser amounts could also be produced. Where the range of hills extends across the direction of flow much of the moisture in the air is condensed and rained out on the windward slopes, leaving the leeward land sheltered and dry. This rain shadow effect is of very considerable importance in many areas of the world, where high mountain ranges lie across the path of the prevailing winds. A further important consequence of hilly regions is that its slopes will vary in steepness, and this becomes all the more important as one moves to

The snowline in the Alps above Zermatt in Switzerland in June. The temperature must still be below 3°C for the snow and ice to persist.

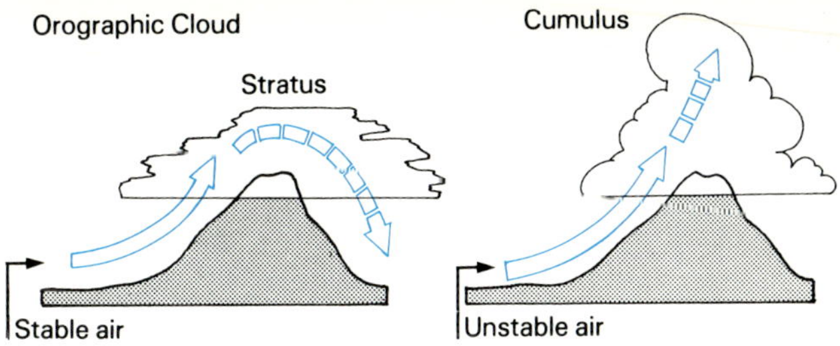

Fig. 27 An air flow crossing mountains may give rise to orographic cloud, which under stable conditions will be stratus (*left*). Unstable air on the other hand may form cumulus clouds of very great extent (*right*).

Orographic stratus forming over the coastline in a wind from the left. The thinning of the layer of cloud down-wind over the land can be readily seen.

higher latitudes. Ground which slopes upwards towards the Sun by too great an angle may receive no direct winter sunshine at all.

Sunslopes – which in the northern hemisphere slope upwards from south to north and vice versa in the southern hemisphere – always receive more solar heating per unit area than anywhere else and naturally this radiant heating reaches its maximum at the summer solstice. However the higher midday temperatures found over such sunward slopes can cause strong thermal currents at many periods of the year that can readily spark off cumulus development and eventually lead to a shower.

Orographic cumulus clouds which had built up over the Cullins in the Islands of Skye in Western Scotland are beginning to clear away and reveal the tops of the hills.

When the airflow is bubbly but still not producing showers, a ridge of high ground often provides just that extra bit of forced uplift that is the final straw and initiates a shower in the region of the hill. So when showers are expected to be few and far between, one of the places most at risk is close by a ridge of hills. When showers are expected to be frequent in most areas, the additional boost provided by the hill when its sunslopes face the prevailing wind may well make the difference between a shower and a thunderstorm or cloudburst. Higher sunslope temperatures combined with low pressure-induced surface convergence and accelerating forced (orographic) uplift of air, generating very strong upcurrents into already unstable and moist air, are just the requirements for spectacular cloud growth, in which the water droplets grow so large and numerous that they fall as the exceptionally heavy shower known as a cloudburst.

In winter the freezing level of the hillside air is much closer to sea level air than in summer. Often in winter the air on exposed hills will be freezing, while only a hundred metres (or yards) lower down over flat land the air temperature remains above freezing. This small difference in temperature makes all the difference to the climate of high ground. When the snow clouds come, the freezing temperatures mean that the snow cannot possibly evaporate as it falls on to the higher ground, whereas over low ground the fall may be melting or turn to rain. The snow line (the height above which snow may be expected to fall) varies throughout the winter, but should not be expected as low down as places that have air temperatures in excess of 3°C (37°F).

Fig. 28 Moist air rising over mountains deposits rain on the windward side of the range. When it sinks on the leeward side it warms at the dry adiabatic rate as a föhn wind.

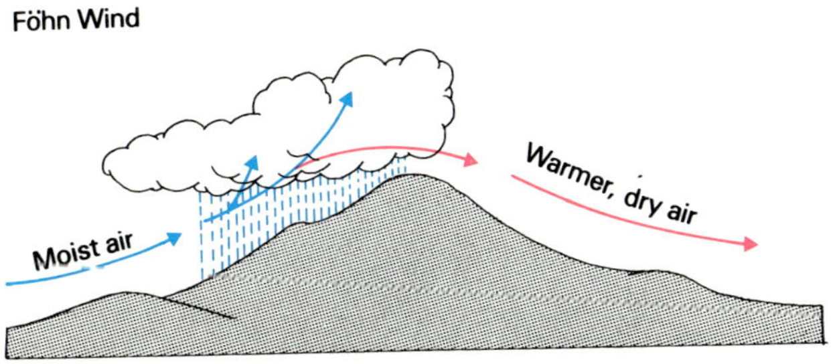

As air descends on the leeward side of any high ground it grows warmer; if it has deposited any of its moisture over the hills it will be warmer and drier than at the same level on the other side. This **föhn** effect can also occur when no moisture has been lost, if wave motions force higher level air to descend to leeward of the higher land. Föhn winds can be produced on all scales, but the most extreme cases occur when warm, moist air loses a lot of its water content in crossing a high mountain barrier and when temperature rises of as high as 21°C (70°F) have been noted. The 'Föhn' itself (after which the effect has been named), occurring in the Alps, and the 'Chinook', found on the eastern side of the Rockies, are the most famous examples. Such winds can begin very suddenly, and in winter the rapid temperature rise can trigger avalanches as well as rapidly reducing snow-cover. The low humidity can also produce a substantial fire-risk as the air dries out cut or standing wood and vegetation

Valleys

Where there are hills there are valleys, and these have an altogether different effect on the local weather. Winds will tend to blow around hills and through the gaps between them, bringing their weather to the low-lying regions beyond. This funnelling effect can be very great when the isobars and the valley are aligned and it is this which causes the famous French wind, the Mistral, to be so strong as it comes roaring down the Rhône valley, on its way south.

On clear nights with little wind the air in contact with the valley slopes will cool as the Earth's heat radiates into space. This air is cooler than the air at the same level directly above the valley floor. The cooler air, being denser, runs down the valley slopes and collects as a cold pool over the bottom of the valley; often valley bottoms are moist places anyway because of the presence of rivers or streams, and vegetation. Cool moist air forms mists, thickening during the night into fog, and consequently valleys are major locations of night fog. So cool are valleys at night, in fact, that air frost is at its most frequent there. At times when other places may expect a general minimum night-time temperature of near freezing, valleys will be several degrees below, and of course the ground temperature will be even lower still.

The frost is at its hardest and the fog at its thickest in the early morning, just before dawn. By day, under clear calm skies, the reverse holds true. The morning air in contact with the slope heats up rapidly as the ground absorbs the sunshine. This air is now warmer and therefore lighter than the valley air which is at the same level but away from the ground. In consequence the air over the slopes starts to climb slowly, lifting the fog, which begins to evaporate and thin out.

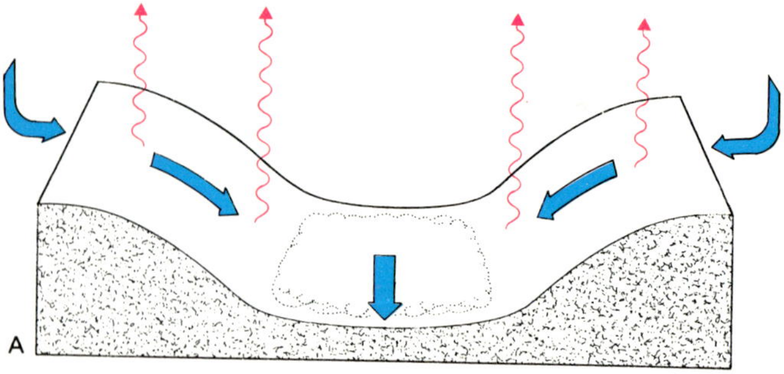

Fig. 29 A valley wind forming during the night-time when radiation to space cools air above the slopes, which then begins to slide downhill. Fog and frost may also be produced in the bottom of the valley.

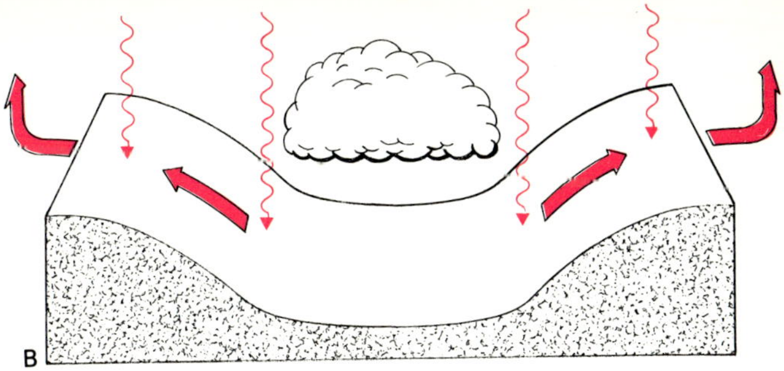

Fig. 30 Conditions in a valley in the morning. Solar heating is causing the fog to rise over the slopes and to disperse.

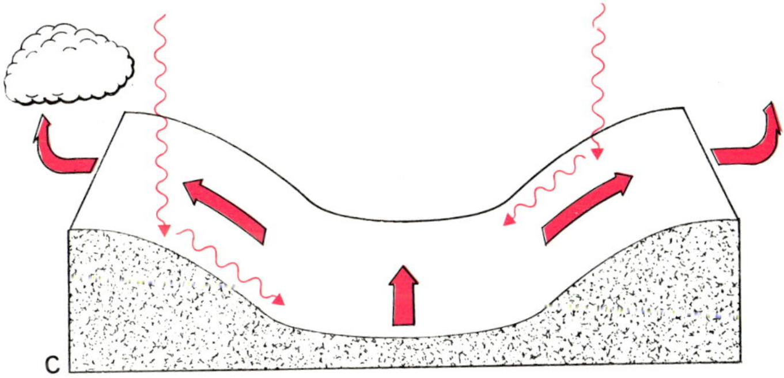

Fig. 31 A fully-developed mountain wind during the daytime. Warm air rising over the sides of the valley produces cumulus clouds over the tops of the hills by late afternoon.

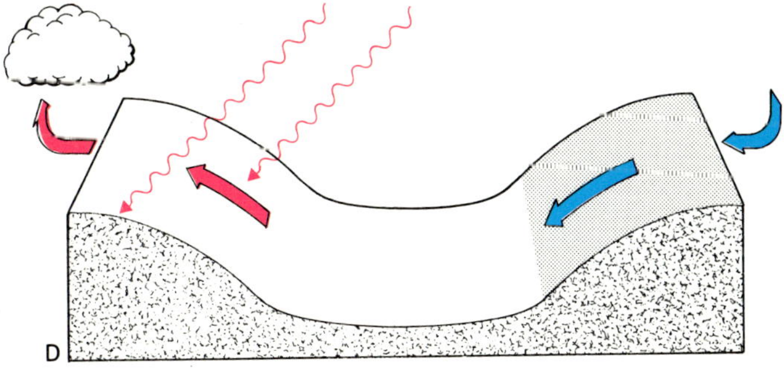

Fig. 32 Cross-valley winds arise when the heating is much greater on one side than on the other. As this frequently happens in narrow Alpine valleys containing lakes, they are sometimes known as lake winds.

During the afternoon the direct and reflected radiant heat concentrated on the bottom of a valley may make it one of the hottest places in the area. For example, if other places may expect a highest temperature of 21°C (70°F), in a valley bottom the temperature could easily soar up to 30°C (86°F). This effect is even greater when the sides of the valley are high above the floor, as they are in many mountainous areas. In these cases, the rising air builds clouds over the surrounding peaks, usually densest towards evening, and it is for this reason that mountaineers are advised to start their climbs early in the morning so that they may reach the summit and begin to descend again before it is shrouded in cloud. When one side of a valley is strongly heated by sunshine, whilst the other remains in shadow, cross valley circulation may build up, with the warmed air flowing up the sunslope and possibly giving rise to cloud at the top.

If the temperature differences and the actual ground slopes are sufficiently great, cold air at night will flow down the length of the valley and out of its mouth over the plains, bringing fog or frost to those areas. This **valley wind** (technically known as a form of **katabatic** wind) has its counterpart in a **mountain wind** flowing up the valley (**anabatic**), during the day. Similar katabatic **fall winds** are produced by the extensive cooling of the air above a high plateau or ice- or snow-covered area and are naturally particularly notable in winter. The

movement of pressure systems frequently is the trigger which causes the mass of cold air to come cascading down the slopes (*Fig. 33*). Despite warming adiabatically, fall winds remain colder than the lower air which they displace. Where the outlets from the high region are restricted, strong valley winds can be produced, and particularly extreme examples are found on the edges of the Greenland and Antarctic icecaps where winds may blow for many days at well over 150 kph (100 mph).

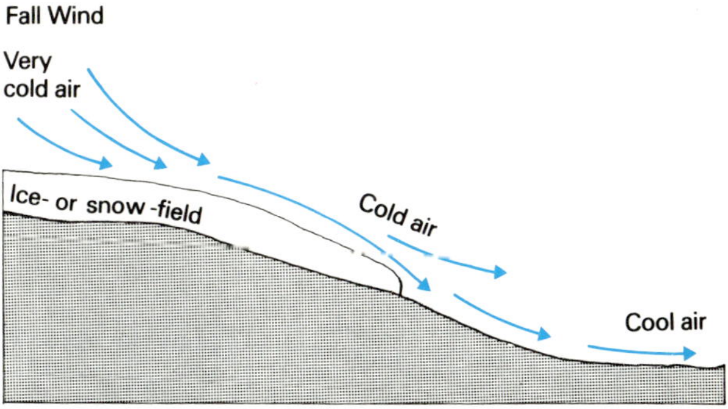

Fig. 33 The formation of katabatic (or fall) winds. Very cold air built up over any high area may cascade down over lower regions, and despite warming with descent, remain colder than the surrounding air.

This valley fog had formed overnight beneath an inversion near Termignon in France, and was just beginning to break up and disperse when this photograph was taken in the morning.

Coastal regions

An important form of microclimate dominates the weather over coastal regions. The sea temperature over the year changes relatively little and slowly due to the high thermal capacity of the water. By contrast the land in summer heats up rapidly by day, so that the air inland becomes buoyant, less dense and therefore of lower pressure than the air over the sea. Low-level air moves from high to low pressure in the form of a cool onshore **sea breeze** (*Fig. 34*), damping convection so that coastal areas are typically cloud-free. They also benefit from additional reflected sunlight from the water. Further inland the leading edge of the sea breeze front may produce showers as well as convective clouds. At night the whole process is reversed. The land cools down to a temperature below that of the sea, producing gentle and pleasantly mild offshore breezes.

If the initial convection which started the whole sea breeze circulation is inhibited because the air is far too stable (which it might be if, say, it is of tropical origin), then there will be no sea breeze during the day. Also if the prevailing wind is offshore during the morning, the sea breeze may not become strong enough to overcome that wind.

In winter probably the most important aspect of coastal weather is the relative warmth of the sea, which heats the onshore winds and protects the coastal regions from the freezing temperatures that may occur further inland.

During winter snowfalls, the coastal regions often escape with just melting snow or rain. Also in winter the relatively warm seas produce weak damp thermals that only have to rise a short distance into the cold air above before becoming saturated. Thus low sheets of dull grey cloud are formed and are a feature of onshore winds during winter.

Fig. 34 The daytime sea breeze. The circulation may cause a sea-breeze front with convective cloud to move far inland during the day.

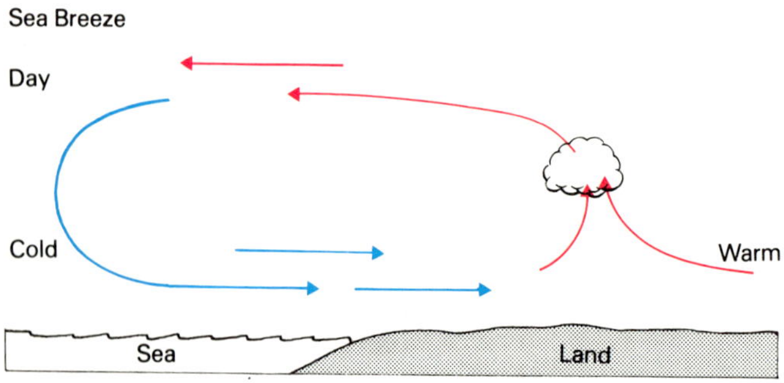

In this picture of southern Norway, with a southerly wind (blowing from the bottom), the coastal strip is clear, but cumulus are forming further inland.

A Gemini photograph of the San Francisco Bay area of California shows strikingly how the land is clear of any cloud whilst an altocumulus sheet covers the sea.

A satellite picture of the north-eastern Pacific with layer clouds over the Aleutians, mixed clouds over Washington and Oregon, and sea fog off the coast of California.

They may push far inland, causing gloomy weather, hill fog and drizzle. Somewhat similar cloud may form when warm moist air is uplifted over sea cliffs or hills, although such cloud is usually localized, and often disappears later in the day.

When the air flow is warm (for example, if it is of tropical origin) it is noticeable how cumulus will form over the warm winter seas but not inland, whereas in summer cumulus forms readily over the land while over the seas the skies are clear blue.

Generally the sea is still at its coldest during spring, and so cools the air in contact with it. When the overlying air is slow moving, the cooling can be enough for water vapour to condense, forming hazardous **sea fogs** that frequently invade coastal areas during the day. Weak sea breezes and turbulence tend to thin the fog by day, but at night when the wind drops, the fog thickens and extends inland.

Few people live out at sea, which is just as well, for worse things really do happen at sea. The uninterrupted surface means that there is little frictional drag to slow down the winds, so they are stronger. The constant evaporation fuels the storm clouds, which are made self-perpetuating by the release of latent heat of condensation. As a storm brews, the reduction of atmospheric pressure allows the sea to rise, creating a storm surge of exceptionally high waves. When such low pressure approaches or crosses coastal regions and rivers running into the sea, the surge of high water may be driven onto the land by onshore winds. Flooding may be even worse if the storm coincides with spring tides when the gravitational effects of the Sun and Moon also combine to raise the water level, and also if the land is already flooded with water from the heavy rain common to the centre of depressions.

A bank of sea fog is visible on the right with low stratus over the downland on the left. The mock sun shows the presence of ice crystals in the higher clouds above.

Storm conditions affecting the Brent B oil rig in the North Sea. Waves reached 23 m (nearly 75 ft) above 'sea level' in 160 kph (100 mph) winds.

Fog

A **fog** is defined (by international agreement) as occurring when visibility is reduced below 1 km (1094 yards) by water droplets in the air: the term **mist** is sometimes used when visibility is greater than 1 km (1094 yards). It is only when visibility is reduced to less than 200 metres (650 ft) or so that ground transport is seriously affected. Dry hazes are dealt with elsewhere (*p. 50*).

Fog forms when moisture-laden air is cooled so that condensation occurs. The principle is the same as that which causes clouds to form (*p. 30*), the only difference being that fog occurs at ground level rather than high above it.

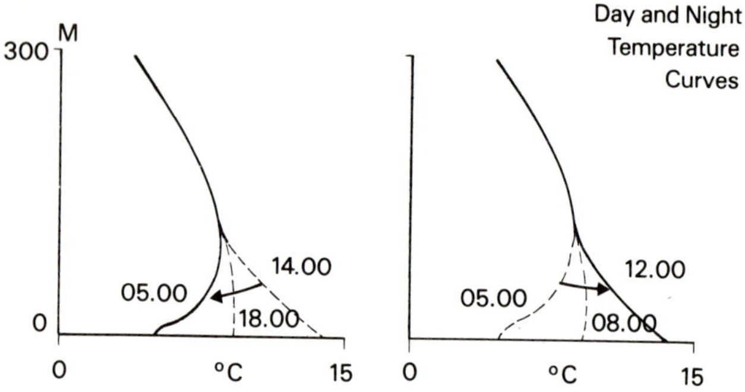

Fig. 35 Typical temperature curves for the lowermost layer of the air during the evening and night (*left*), and from dawn onward (*right*).

The process can occur in various circumstances. Air that is forced to climb a hillside will become cooler and cooler until eventually it is so cold that the water vapour it contains condenses into water droplets – the air is saturated – and **hill** or **upslope fog** forms. Alternatively, air that is forced to move over surfaces that are much colder than the air itself – surfaces such as cold rivers and seas, or snow-covered ground – will be cooled, and if cooled enough will become what is technically known as **advection fog**. But the most usual process by which fog forms occurs on still clear nights. The earth radiates its heat into space and so the ground cools. The air in contact with the ground is also cooled, becoming colder than the air above, so that it is unable to rise at all. Instead the chilled air begins to be very humid and, as long as no wind arrives to stir up the air, mist forms and eventually thickens into a **radiation fog**. Occasionally both processes may operate at the same time. Dew may be regarded as a radiation fog which actually forms on objects rather than in the air above.

Therefore the probability that fog will form is at its highest at night under starry skies when the wind is calm and the air is of high humidity. To predict fog, watch for the following pointers in the evening: the barometer points to the high part of its scale; the temperature is falling rapidly; the air should be of tropical origin; the evening wind should be dying away completely.

Fog forms soonest and thickens most rapidly in valley bottoms and by riversides and thick vegetation. The last places to become foggy will be the highest ground and town centres. If there is fog in either of these places it is likely to be widespread elsewhere. Fog is densest and most extensive at dawn after the long night-time cooling of the Earth's surface.

In towns, any smoke injected into foggy air thickens into **smog**. Sulphur dioxide in the smoke reacts with moisture and oxygen in the fog to form sulphuric acid, which corrodes stonework and attacks the

Fig. 36 Advection fog (*left*) is produced when air moves across a colder surface. Radiation fog (*right*) forms when both surface and air above cool by radiation to space.

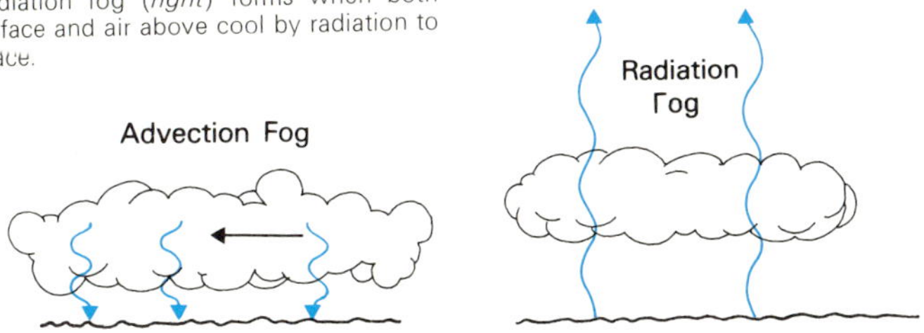

The remnants of overnight radiation fog lifting and dispersing in the Yosemite Valley. Denser patches in the background below El Capitan are disappearing with the heat of the Sun.

respiratory system. Exhaust and industrial fumes can add to this, of course, although they are also involved in other forms of pollution hazes (*p. 84*).

If during the morning some heat from the sun manages to warm the ground and begins to raise the temperature of the fog, more and more of the fog droplets will evaporate; the fog will then start to thin. Signs revealing that fog is about to clear are: a significant brightening in the early morning; the wind beginning to rise; and patches of sky just visible through the fog directly overhead. The fog will then lift slowly and form low grey sheets of cloud, which, if lifted before noon, will most likely break up and disperse. The thicker the fog the longer it lasts and on some days in winter it will persist all day. The sun just does not have enough heating power to shift the denser fogs in valleys and by rivers. Signs which reveal that fog is going to persist all day are: no significant brightening during the late morning; a lack of wind; and the air remaining heavy with moisture.

Over estuaries and coastal regions in the period between late winter and early summer, fog often lasts all day. **Sea fog** is dense and pervasive, pushing inland when the prevailing light wind is onshore, and lasting just as long as the warm, moist and stable stream of air continues to be cooled to saturation by the underlying cooler sea. But strong winds will break up a fog – the turbulence and continued stirring of the air warm it and disperse the moisture, thinning the fog so that it eventually dissipates entirely. Along coasts night breezes will often carry the sea fog back out to sea. Since the amounts of moisture in the air vary from place to place, fog is often patchy, especially early in the night.

On occasions when cold air passes over warm water (rather than the temperatures being reversed), vapour from the water condenses in the air to give a **steam fog** – this is frequently seen over wet roads heated by the sun. The same process occurs on a large scale in arctic regions to form **arctic sea smoke**, but because the air above is so cold and dry, the bottom layer is very unstable and the height of the sea smoke is usually less than 10 metres (35 ft).

It is possible for fog droplets to exist in supercooled form, and when such a **freezing fog** flows past any objects the droplets freeze immediately to give a deposit of **rime** on windward surfaces. This is opaque, contains a lot of trapped air, but is not crystalline, unlike hoar frost (*p. 78*). At even lower temperatures an **ice fog** may form when water droplets turn into ice crystals. It is not uncommon where vapour is released into very cold air at −30°C (−22°F) or less, and may be formed by aircraft, above towns (or even by herds of caribou or reindeer). Natural ice fogs are of lesser density and contain fewer of the ice crystals, which glittering in the Sun, give rise to the popular name of 'diamond dust'.

This picture makes the restricted depth of some layers of fog (which can also be described as low stratus) quite apparent. Nevertheless ground-level visibility is obviously low.

Arctic sea smoke in cold air above a relatively warm sea surface. The restricted depth of the layer of this steam fog is evident in this photograph.

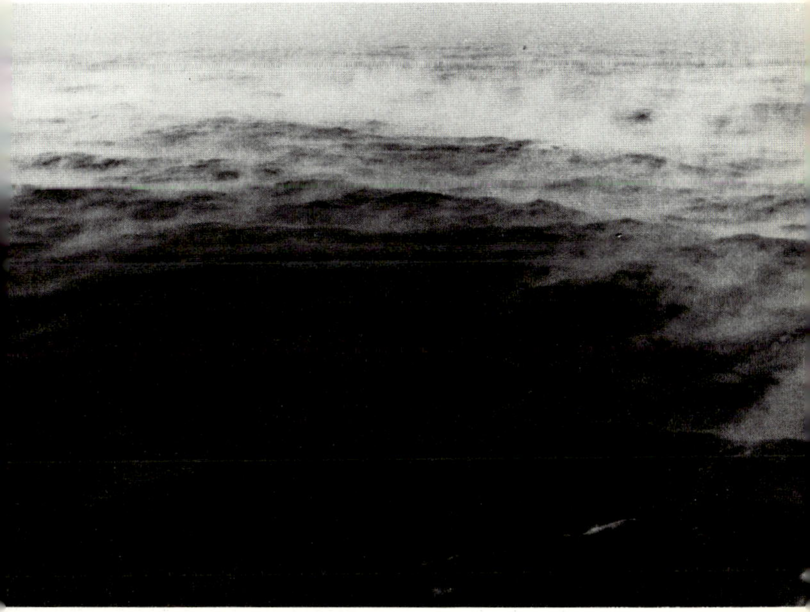

Frost

Although many areas of the world have fairly predictable yearly temperature patterns with long periods when it is below freezing for the whole day, other regions may have more variable weather, or only experience frost on rare occasions. The prediction of such frosts can be of vital importance to fruit growers, farmers and gardeners because of the damage it can do, and of course very low temperatures can be very serious to a community, especially if such conditions occur only rarely.

Most people think of frost as the icy white deposit on the ground and on windows but strictly speaking **air frost** occurs when the temperature of the air 1·2 metres (4 ft) above the ground falls below 0°C (32°F); **hoar frost** or **ground frost** occurs when the ground itself is freezing.

The appearance of fog as the night temperature drops means that the rate of fall of temperature will be slowed, so that frost may well not occur. When the air is very dry, however, frost will form as the temperature falls below 0°C (32°F) and if the fall is sustained so that eventually what little moisture there is in the air condenses, **freezing fog** will form. Freezing fog is particularly hazardous because it is made up of supercooled water droplets that will immediately freeze on contact with a solid object to give a deposit of rime (*p. 76*). As this is opaque it is particularly hazardous to drivers.

When there is no fog and temperatures fall below freezing, not only do water droplets freeze on to windows and other objects at tempera-

The hoar-frost in this photograph is in the form of delicate crystals which formed on cooled surfaces overnight. A small thermometer screen is to the left.

tures below freezing, but ice crystals and supercooled water vapour also collect on freezing objects and contrive to create one of nature's most delicate pictures, with fans and needles of **hoar frost** on every surface. When water or partially melted snow that is already lying on the ground freezes it forms what is known as **ground ice**. But freezing rain – mentioned earlier (*p. 48*) – frequently forms when rain-bearing layer clouds advance over a very cold land surface covered by a shallow layer of air below 0°C (32°F). Raindrops hitting the ground have time to form a film of water before they freeze into a clear sheet of ice. Such a **glaze** or **glazed frost** can cause severe damage to trees and other plants as well as occasionally damaging power and telephone lines through the weight of ice. On road surfaces it will often form a thin sheet of ice which is nearly transparent and therefore appears dark and this is commonly called **black ice**. A road covered with black ice looks exactly like an ordinary wet road, and motorists may be misled into thinking that conditions are not hazardous. In fact, black ice is so slippery that it is difficult even to walk on; driving is almost impossible and is certainly dangerous.

The fact that the temperature is below freezing does not necessarily mean that everything will be covered with a white frosty deposit. On nights when the air is too dry no such deposits are possible even though everything – the ground and the air – is freezing cold.

Predicting frost is not difficult; in fact it can be done with consistent success. The probability of frost is at its highest when the day's air mass is of polar origin, when the wind is light or calm, and when the skies are fairly free from low cloud. Under these conditions the fall of air temperature which always occurs at night will be at its most severe. The amount of fall can be estimated by the method given on *p. 150* which involves taking two temperature readings an hour apart in the evening, and from these estimating the decline by dawn at which time air temperature is lowest (*Fig. 35*). Provided the skies remain clear this method is fairly accurate.

Frost is affected by wind speed, since a strong wind will cause the cold night-time air to become mixed with the warmer air above, so holding off the frost a little longer. However, in freezing conditions a strong wind feels decidedly colder than a light wind because of its penetrating effect, or 'wind chill factor'. A relatively mild air frost in a strong wind can therefore be classified as a severe frost even though the thermometer seems to be at odds with this description; in light winds the temperature needs to fall below −5°C (23°F) before the frost is classified as severe.

The ground will generally reach freezing before the air, so on nights when the air temperature hovers above freezing the ground is nevertheless frozen. The expansion of freezing water trapped in the top soil is sometimes enough to uproot plants even though the air is still above freezing.

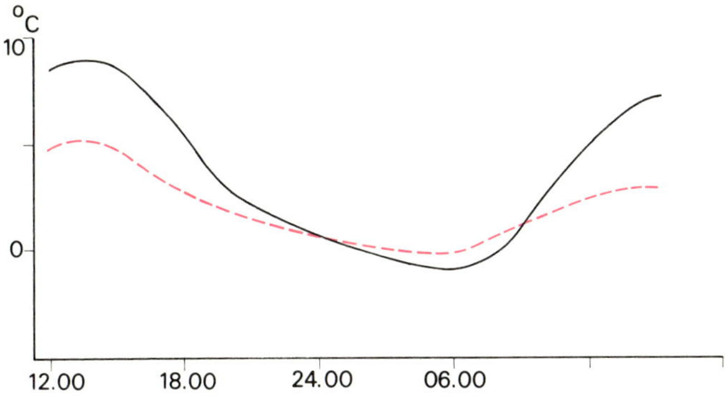

Fig. 37 A typical diurnal temperature curve. The range under overcast skies is less (dashed line) than when the sky is clear (solid line). With snow or ice cover the lowest temperature is reached shortly after sunset.

Rough crystals of rime which have been deposited from supercooled water droplets which froze on contact with some surfaces to give a partial covering.

Soil and vegetation

To a certain degree the types of soil and vegetation in any locality have an influence on the weather. The overall colour and the amount of air and water trapped in the surface layers determine the degree to which the surface heats up by day and cools by night. Surfaces which are light in colour are highly reflective and absorb little of the Sun's heat – thus brilliant snow is slow to melt. Dark soil absorbs more heat, and so bare dark soils are the last to be affected by night frosts.

Air is, of course, an efficient insulator, and this means that any air trapped in the soil surface lessens the effect of daytime heating on the layers below. Thus sand is very cold deep down even though the surface can become searingly hot. It also loses its heat rapidly at night because the trapped air does not hold its heat as efficiently as water.

Water trapped in the soil reduces the amount of both the daytime heating and the night-time cooling. Thus during the day, wet soils show little variation in temperature. Both wet soils and vegetation evaporate moisture into the air, producing a considerable cooling effect on the surface. As a result, frosts and dews first form on the lawns of gardens, and shallow mists appear first over the grassy plains. Along a road at night patches of fog are found where the nearby vegetation is thick and the soil very moist, even if the road is level. In the clearances between patches of fog the nearby soil will be found to be relatively bare and to contain far smaller quantities of trapped water.

Another effect of vegetation which is of great importance to many persons is its production of **pollen**. Although hay fever sufferers can be allergic to pollen from many plants, that from grasses is frequently the most irritating. The amount of grass pollen released into the air is highest during early summer when the sun warms the grasslands on days when the air is unstable. The convective thermals carry the pollen up into the stronger winds aloft where it is carried over the surrounding areas, remaining suspended until convection dies down, typically around five or six o'clock in the afternoon. This is the peak time for pollen when the daily pollen count reaches its maximum. The highest pollen counts are in high summer when moderate winds often bring dry weather. The lowest layers of air are unstable, due to the high surface temperatures, so that convection readily disperses the pollen.

As rain washes pollen from the air, it could be said that the best environment for hay fever sufferers would be out at sea in the pouring rain; but failing that coastal areas are generally free from pollen because of the onshore pollen-free sea breezes. Depending upon the amount of convection and whether any inversions are present, higher altitudes will also tend to have clearer air.

The effects of vegetation are evident in this photograph of an Alpine valley, where the moisture-laden air over the tree-clad slopes produces persistent cloud/fog patches.

Towns, cities and pollution

The lie of the land affects the local weather in various ways; so too does man. The effects of his presence are most noticeable over large towns and cities, where an urban climate has produced a **heat island** which interferes with the free exchange of heat, moisture and momentum. During the day the concrete jungle downtown absorbs the solar heating due to its high thermal capacity, re-radiating the heat slowly later like a giant night-storage heater. Hence at night cities are warmer than the surrounding district by up to 5°C (9°F). It is for this reason that the incidence of fogs and frosts is at its lowest in cities.

By day, all other things being equal, the city is a source of strong thermals arising from the excess of heat energy released into the atmosphere. The higher temperatures and thermal currents create an artificial heat low – lower pressure due to hot air rising. The air surrounding the city is therefore at slightly higher pressure and starts to converge on the city to take the place of the rising air, establishing a weak circulation. When the air is unstable and moist, local town showers may develop. On the other hand when there is a definite 'lid' to general convection in the form of layered cloud, the increased turbulence over the city may be just enough to break up the cloud and provide a hot and sunny microclimate.

Winds in the city are rather curious. Clearly, the more walls there are the more shelter there is from the wind; this means that courtyards and gardens, for example, may enjoy a good deal of calm, while out in the country the wind is still quite strong. On the other hand, the lines of buildings tend to funnel the wind along the streets causing it to accelerate through smaller and smaller gaps, so that at a crossroads in a city the wind may gust to gale force and swirl up all the litter of the streets, while out in the country it is still only a breeze. Similarly, the obstacles presented by large buildings may lead to considerable eddies, and consequently produce violent gusts of wind from apparently contradictory directions.

The type of pollution experienced in high summer in the city can be an oppressively hot affair. When pressure is high overhead, so that the upper air is warm and sinking, the lowest layers of air do not mix with, and are thus no longer diluted by those layers of air higher up. Smoke from chimney stacks then emerges almost horizontally, demonstrating the stability of the surface air. Gentle eddies in the air gradually pull the pollution down to street level. The concentrated burning of fossil fuels injects more and more carbon dioxide, sulphur dioxide and dust into the surface layers of air, all of which is trapped by the prevailing high pressure, and so the air quality falls. The heat of the ground becomes trapped by the pollution, and the temperature rises still further, while the sky turns a hazy, sometimes dazzling, milky-white, and visibility becomes poor. This type of weather usually breaks down when the temperatures rise so high that the convective lid is broken and the

In most developing countries pollution control is an expensive luxury which is not yet practical. This suburb of New Delhi in India shows conditions which are repeated worldwide.

resulting heavy shower or thunderstorm washes away the pollution.

A different form of pollution may occur in damp weather. Then **smoke fog** or **smog** may be formed by the waste products from coal being burnt domestically and industrially finding their way into the fog and becoming chemically active. The sulphur dioxide released in the smoke reacts with water to produce sulphuric acid. This pollutant slowly corrodes buildings, and attacks the respiratory system in human beings. Prolonged inhalation of wet smog (fog plus sulphur dioxide) is very likely to cause bronchitis. Levels of sulphur dioxide found in cities are commonly some thirty times those found in the natural environment, where decaying vegetation is the source.

Sulphur dioxide stays in the atmosphere for a relatively short time – only two to four days – so most of it is deposited fairly close to its source. Many cities are situated in valleys which are prone to fog, but with more stringent controls, many now experience less pollution than formerly. The infamous London smogs or 'pea soupers' for example, are now very rare as a result of legislation which initiated the use of alternative fuels and required industrial chimneys to be built higher, and so become more effective in dispersing their pollution into the less stable air above the fog.

But just as the incidence of wet smogs has decreased, so an entirely

new type of smog has recently been observed and is on the increase. Unrelated to fog, this modern man-made smog is formed under sunny skies such as those of California, when thermal convection is prohibited by a surface level inversion of temperature resulting from strong subsidence of air and high atmospheric pressure. The cause is the automobile. Hydrocarbons and oxides of nitrogen which are released in the exhausts from the internal combustion engine accumulate under the weather conditions just described and the oxides of nitrogen react with sunlight to produce ozone. The ozone reacts with the hydrocarbons to produce a suffocating smell and the hitherto unknown peroxy-acetyl nitrates, which among other things are a health hazard, irritating the eyes and nose. **Photochemical smog**, as it is known, hangs on hot days as a pall over urban areas of high traffic density. The sticky haze is evidence of high levels of oxidants being produced photochemically at ground level. Ozone, the chief oxidant, can be found in concentrations five times as high as the natural background level and the photochemical reaction is largely responsible for the oxidation of sulphur dioxide producing abnormally high levels of sulphuric acid and sulphates. The residence time in the atmosphere of ozone and sulphuric acid produced photochemically

Above Smoke plumes rising strongly in unstable air. Such conditions are accompanied by considerable mixing so that pollution is diluted – and spread over a larger area.

Left Pollution is being trapped under an inversion in this long-vanished scene once common in the towns of the Potteries in England. Changes in production methods and legislation have produced cleaner air and even the famous 'bottle' kilns have vanished.

near the ground is only a few hours, and so the hazard is usually a temporary one during daylight hours. However the stringent regulations which are enforced in some areas regarding the design of automobile engines and exhaust systems, has gone a good way to controlling this dangerous pollution.

Apart from photochemical smog and the noise pollution produced by high density traffic, there is yet another atmospheric pollutant emitted in the exhausts from the internal combustion engine and which is the subject of much debate. Lead is consumed at something like 0·08 grammes per mile by an 'average' car and around a quarter of this is thrown out into the atmosphere in such a state as to remain airborne for up to a month. Typical concentrations of lead found in cities are one hundred times that found occurring in the natural atmosphere. Lead can accumulate in human beings and ultimately affect the central nervous system, young children being particularly at risk. It is now widely accepted that there is a correlation between high lead levels in the body and mental disability. However, again, now that the danger has been recognised, many states and countries either have cut or are reducing the level of lead in vehicle fuel, and some including the U.S.A. are eliminating it completely.

The depression sequence

The two processes by which raindrops are formed within cloud have already been described as have some of the varieties of cloud associated with them. However it may be more useful to consider the two types of sequence which lead to extensive falls of rain: the frontal depression sequence described here, and the shower and thunderstorm sequence (*pp. 98–101*).

Because of the distinct structure in a low pressure system, the clouds, rainbelts and winds follow a definite sequence. Obviously the position of the observer relative to the centre of the low makes a considerable difference, but it should also be remembered that all depressions are different, cloud and frontal types may vary within one depression, and that frequently the characteristics may be weakened and some cloud types missing – this particularly applies to the high clouds.

Before the warm front

As the weather associated with depressions is of such importance it is worthwhile to consider it in some detail, concentrating on the 'classic' sequence found at ana warm and cold fronts, at which the warm air is rising. In this case forecasting the arrival of the belt of warm front rain is reasonably straightforward. It is a common misconception that cirrus – the wispy ice clouds at the highest level – are invariably the forerunners of rain. In fact it is only when the cirrus continues to invade and thicken across the whole sky that rain should be expected to follow. The presence of cirrus indicates that the high altitude air is saturated. By watching any exhaust trails from aircraft at cirrus altitudes the degree of saturation of the high-altitude air can be estimated. Condensation trails, or contrails, that persist indicate that the air is very moist, and when thin veils of cirrus are also present the trails will cause the cirrus to thicken. The cirrus thickens and lowers in the direction from which the rain is approaching, and a halo will often form around the Sun (or Moon). If this sequence is rapid, rain may be expected to follow within a few hours. If the sequence is slow, taking most of the day to happen, the rain is likely to follow a day or two later. With an average speed of advance of 50 kph (31 mph) rain may be expected about 9–10 hours after the cirrus is noticed thickening overhead, and the front itself about 6 hours later still. Ahead of this approaching warm front the clouds are likely to be cumulus, but as the higher clouds encroach the cumulus become flattened and shallower, perhaps changing into stratocumulus, eventually dying away entirely as the convection is reduced.

An important clue to the exact development of the weather lies in the direction of the winds at various heights. For the present we will assume that the movement of the depression will carry both warm and cold fronts over the observer with a distinct warm sector in between

The first indications of a warm front approaching from bottom left. The cirrus cover is increasing and the aircraft contrails begin to persist.

Fig. 38 The structure of a depression. The air has cyclonic (anti-clockwise) circulation at the surface, but the rising warm air flows out anticyclonically at height. The whole system moves in the general direction of the jet stream.

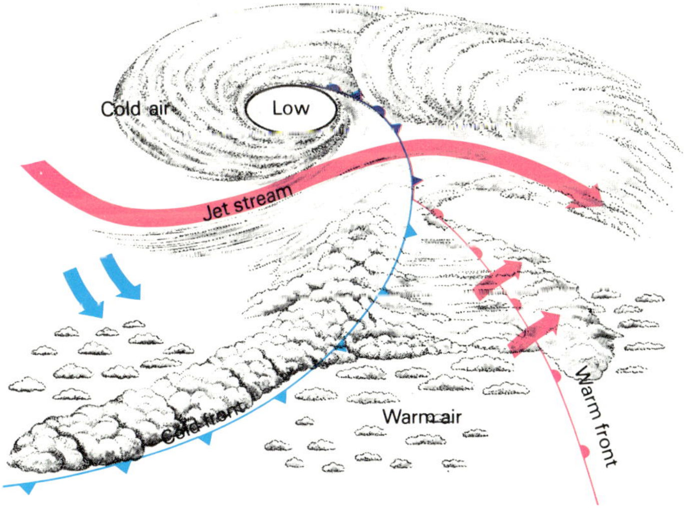

Left A veil of cirrostratus now covers the whole sky. For a short while a solar halo is visible before the high cloud cover becomes even denser.

Opposite A watery sun showing through fairly thin altostratus. Soon even this will be blotted out by the ever-increasing thickness of the clouds.

(*Fig. 38*). Under these circumstances (and in the northern hemisphere) an observer facing the front will find the low-level clouds moving from left to right, the middle-level clouds moving more or less directly towards him, and the highest possibly even right to left, that is, parallel to the front. The winds therefore **veer** (change clockwise) with height, so that crossed low and high winds with this wind-shift area classic symptom of approaching warm air and deteriorating weather. (It may be noted here that the converse situation when winds **back**, that is change anticlockwise, with height usually implies that cold air is becoming more dominant and the weather is improving.)

The rainbelt

Reverting to our sequence of clouds as the rain approaches, the Sun and halo are lost from view and the sky darkens as it fills with layered cloud with a progressively lower base. The rain falls from nimbostratus clouds (*p. 48*): dark and featureless clouds with ragged bases sometimes so low that high ground and the tops of tall buildings are completely lost in them. Provided the wind is still at least breezy, the rain belt will move through within three or four hours; but where a calm accompanies the rain the duration may be much longer. The continuous rain may sometimes be accompanied by much heavier showers where instability and convection occurs within the warm air aloft. Of course in winter freezing rain or snow may precede the rain, or if temperatures are low enough the whole precipitation may be in the form of snow.

The warm front

Throughout these events the pressure has been dropping as the centre of the low moves closer to the observer, but at the warm front it steadies at its lowest point. The surface wind which has backed slightly, been strengthening continuously and perhaps become gusty, now veers with the passage of the front and swings from south to south-west or west. The continuous rain moderates and peters out, the warmer surface air arrives and the clouds begin to brighten or break up.

Fig. 39 A typical ana warm front. The tropopause is higher over the warm air mass, and there is frequently a break in its level at the upper junction of the air masses, where the jet stream is located.

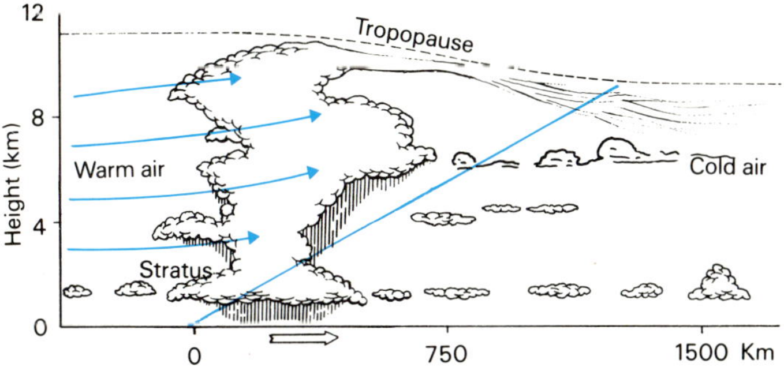

It is, of course, the warm air riding up over the cold air on a slope of, typically, 1:100 or 1:150, that has produced the rain and the sequence of high, medium and low clouds. The warmer air is usually of tropical origin, relatively moist, easily forming low sheets of cloud.

The warm sector

The proximity of the parent centre of low pressure determines whether the warmer air will be dull and drizzly or well broken up and fine. If the parent low is hundreds of kilometres or miles away, warm sector weather often produces some of the most vivid skies with low, medium and high level clouds brilliantly lit by sunshine, and the weather is comfortable because of the warm breeze (although in summer over strongly heated land, convective showers and thunderstorms frequently occur). If the parent low passes close by, then the warm front rain will continue on and off until the colder air of polar origin finally returns to freshen and clear the skies.

The cold front

The cold front is against the flow direction, and has a much steeper slope than the warm front, very typically 1:50 to 1:75. The associated rain is therefore of shorter duration, frequently very heavy and accompanied by thunder. In some regions however, the warm sector air is commonly stable, and nimbostratus is the main rain-bearing cloud form. With the passage of the cold front the wind suddenly veers again (perhaps from south-west or west to north-west or north). In addition the temperature falls sharply and the pressure rises. The weather behind the cold front depends strongly upon the stability; when the air is unstable, perhaps warmed over the sea, strong cumulonimbus may build up. Often, compensating sinking air follows the passage of the cold front giving clear blue skies.

Fig. 40 An ana cold front. Just as for the warm front, a jet stream may exist at the change in the level of the tropopause. The extent is much less than that of warm fronts.

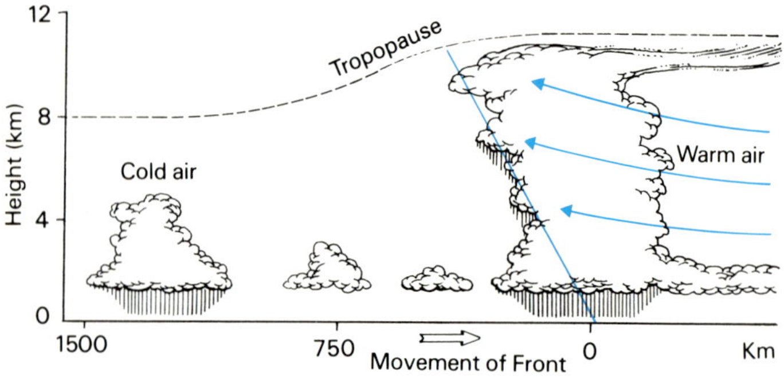

Occluded fronts

What of occluded fronts? They will show most of the distinct features of both warm and cold fronts, but the main difference is the absence of the warm sector, so that the cumulonimbus follows immediately upon the warm front sequence. There is a sharp pronounced, wind veer (up to 45°), and a drastic drop in temperature with the passing of a cold occlusion (*Fig. 42*). The less frequent, warm occlusions are usually less distinct and shorter-lived.

Fig. 41 Occluded fronts. A cold occlusion **a** has cool air preceding colder air at the surface. A warm occlusion **b** has warmer air following. In both the warmest air has been lifted from the surface.

Fig. 42 A representation of a cold occlusion. In effect warm and cold fronts are combined, so that the sequence of clouds remains more or less the same.

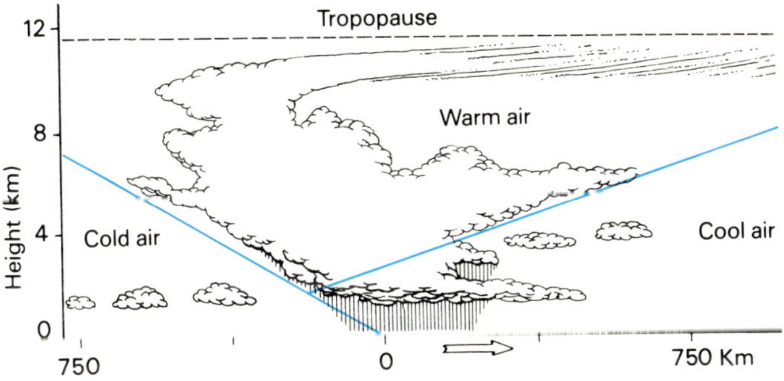

Away from the centre

An observer to the north of a depression centre may note high cirro-stratus overhead sometimes preceded by cirrus and may think that perhaps a warm front is on its way towards the observer, especially as the pressure begins to drop. However on closer examination it will be seen that the surface and higher winds are fairly closely parallel to one another, although actually flowing in opposite directions. The cloud does not build up significantly, but disperses instead and cumulus and other clouds appear while the surface wind backs from south-east through east to north as the centre passes to the south and pressure rises.

On the other side of the depression some cirrus and cirrocumulus may precede altocumulus which never completely covers the sky. Any pressure and wind changes are very slow and high pressure conditions

The main portion of the belt of warm frontal rain has now passed and clearer weather is now approaching. This picture was taken over the foothills of the Alps near Chamonix in France.

A chaotic sky fairly typical of the warm sectors of depressions with low cumulus, altocumulus and cirrocumulus clouds at several levels.

frequently return. In some cases indeed after a distinct warm front has passed, a significant pressure increase caused by the extension of an anticyclonic ridge may completely block the cold front from following.

Isolated fronts

Isolated warm or cold fronts not forming part of a distinct depression system are rather more frequent over continental interiors (such as the U.S.A.) than maritime regions. They show characteristics generally similar to those described, although warm fronts may well have rather greater convective activity and clearer following skies. Isolated cold fronts are notable for the distinct drop in pressure before their arrival and the subsequent rise.

Weak frontal systems

Finally we come to the kata-fronts where the warm air is subsiding. Fortunately these are much simpler, although weaker and less distinct. The subsiding warm air suppresses convection and does not give rise to the extensive high and middle cloud sequence which is such a useful indicator of an approaching ana-warm front. Instead rata-warm front cumulus tends to turn into stratocumulus and this becomes very thick indeed.

Rainfall is light and frequently in the form of drizzle. Behind the front – at which temperature and winds change as at the 'classic' ana-front but in a less marked and slower manner – the stratocumulus thins usually allowing the sky to be seen through small gaps in the cloud cover, finally being replaced in part by stratus. With a kata-cold front the sequence is reversed, the stratus and thin stratocumulus thickening to give light rain, and with cumulus or cumulonimbus in the cold air behind the front, at which the temperature, wind and pressure changes have again been slow and slight. Kata-front depressions are therefore dull, uninteresting affairs, but they are very common in many areas of the world, especially in winter. However as any low pressure area can have a mixture of characteristics along its fronts, it is not surprising that every low pressure system is different, or that they are so important in the weather they generate.

Forecasters tend to err on the side of caution and often predict frontal rain that subsequently does not arrive. The main pointer to look for is a significant dulling over or darkening of the skies, indicating that there is enough cloud depth for the tiny droplets to grow large enough to fall as rain. Another clue to look out for is the speed of movement of the high and medium level clouds. On a day when a rain belt is forecast, if the wind is almost calm and the clouds are merely meandering across the sky, and weak shadows still form on the ground, the rain is unlikely to turn up. This is especially true when the barometer is unwilling to leave the higher half of its scale.

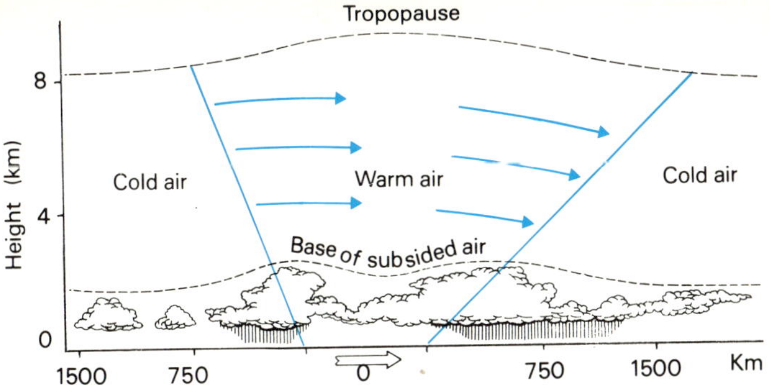

Fig. 43 The subdued structure and clouds of a depression with kata fronts. Thick stratocumulus accompanies both fronts with stratus and stratocumulus in between. The cold air masses are rather warm and dry.

A line of cumulonimbus clouds which were also active thunderstorms mark the position of part of a cold front.

Showers and hail

Apart from the passage of low pressure systems, the other train of events which leads to rainfall is that which, when it progresses to its natural conclusion, is responsible for hail and thunderstorms. What weathermen call a **shower** could be described as a failed thunderstorm, or, alternatively, a thunderstorm could be described as a shower that has got out of control.

The thunderstorm sequence takes place when there are high temperatures in air that becomes progressively colder with altitude. Cumulus grows in the morning sky, indicating that the air is unstable. Since the total amount of convection is revealed by the area covered by clouds, and the amount of balancing sinking air is proportional to the areas of blue sky, showers are only likely to occur when there is a greater area of cloud area than of clear blue sky. Hot spots on the ground, such as towns, help to strengthen the convection; similarly hills will often provide the initial springboard for strong uplift of air and cloud growth. If you have time to stand and watch the sky, developing showers can be picked out as they darken at their bases. They normally affect an area of around 10–12 square kilometres (4–5 square miles) and last for between 10 and 30 minutes. Daytime heating of the land, causing strong thermal upcurrents, is a great incentive for shower clouds to grow and produce bursts of rain. Showers tend to die out at night as convection ceases in the late afternoon and early evening.

When the wind is strong any showers are passing ones, lasting less than half an hour and with a fair amount of sunshine in between. When the wind is weak the duration of each shower is longer, but there are fewer of them during the day.

With showers come blustery winds. These are caused by the violent upcurrents and downcurrents which are always found in cumulonimbus and which affect the air beneath them. The surface wind can quite easily double its strength and reverse direction back and forth as

The edge of this active cumulonimbus shower cloud appears like a series of rings. This structure has been produced by the violent upcurrents and expansion within the cloud.

These four photographs taken on the same day illustrate the formation and growth of potential shower clouds. About 30 minutes separate the first two pictures showing the increase in that short time. The dense, dark bases shown in the third photograph are signs that showers can be expected soon. The final picture shows cumulus and stratocumulus beginning to die away at sunset.

A spectacular example of precipitation from a comparatively isolated shower cloud photographed over the East African plains.

Fig. 44 The currents within cumulonimbus clouds are violent and lead to the circulation and growth of hail. The cloud is frequently preceded by a distinctive arch cloud in the updraught.

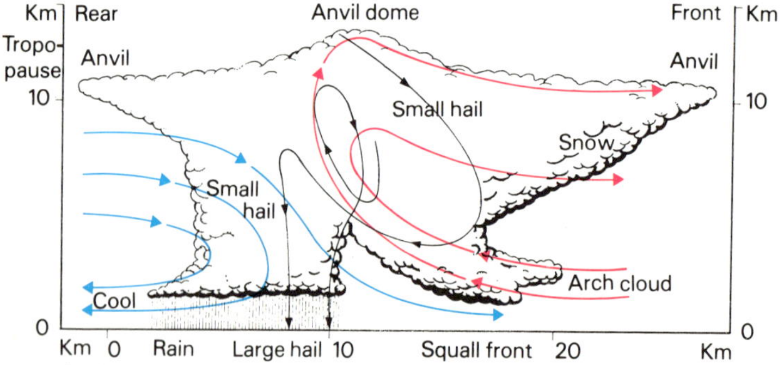

a dark shower cloud passes overhead. Also with a shower comes a drop in temperature, since the falling rain tends to evaporate in the warmer air below producing a cooling effect of, typically, 3°C (5°F).

Showers fall from clouds that are known technically as precipitating convective clouds, or cumulonimbus. As we have seen (*p. 34*) these can grow right up to the stratosphere where the top spreads out to form

an anvil. The largest of all cumulonimbus clouds can grow to some 12 km (8 miles) in height and 16 km (10 miles) across, and readily produce hail, thunder and lightning.

The tremendous up and down currents within a towering cumulonimbus carry ice particles through the whole range of temperatures within the cloud to form **hail**. Small supercooled water droplets freeze instantaneously around ice particles in the upper (coldest) part of the cloud, trapping bubbles of air and adding a glazed layer of ice, while near the lower (warmest) part of the cloud large water droplets freeze slowly on to the ice stone adding a clear layer. In this way hailstones grow larger and heavier as they circulate in the cloud, alternately adding glazed layers and clear layers. Individual stones can grow to a considerable size; the largest stone ever measured anywhere was 19cm (7½in) in diameter, and there is evidence that larger stones have fallen. Hailstones can leave the cloud base at a speed of 160 kph (100 mph). Hail is one of the most destructive weather phenomena, causing terrible damage to crops in regions where other extremes of weather are rare.

Thunderstorms

Thunderstorms require a very strong temperature contrast between surface and air (a steep lapse rate for their formation). This may occur with very high surface temperatures or with unusually cold air aloft – or a combination of both factors. In any case the air has to be extremely moist in depth so that saturation can occur easily at all levels.

There are various possible mechanisms by which the electrical charge may be generated within the cloud. These include the effects of friction between ice particles, and charge separation on freezing of water droplets. The lighter, positively charged ice particles collect at the top of the cloud, and the heavier, negatively charged hailstones and water drops at the bottom. The cloud will discharge to earth when the base approaches a high object on the ground, producing spectacular flashes of forked lightning which heat the air and cause it to expand at supersonic speeds, giving the familiar crashes and rumbles of thunder. The lightning channel is forked because the discharge seeks out the path of least resistance through the intervening air, and this path is usually a tortuous one. The potential difference between the top and bottom of the cloud can also cause a discharge within the cloud itself, which is seen as a diffused glow rather than a flash because its light has to pass through millions of water droplets before it reaches the observer. Since the lightning reaches the observer virtually instantaneously and thunder takes about three seconds to travel one kilometre (five seconds to travel one mile) the interval between flash and the onset of thunder indicates the distance to the electrical storm. When no thunder follows the flash, the storm is probably more than 25–30 km (15–20 miles) away.

Fig. 45 A thunderstorm frequency map for the U.S.A. showing average days with thunderstorms; these are least on the West Coast and most in the Gulf States.

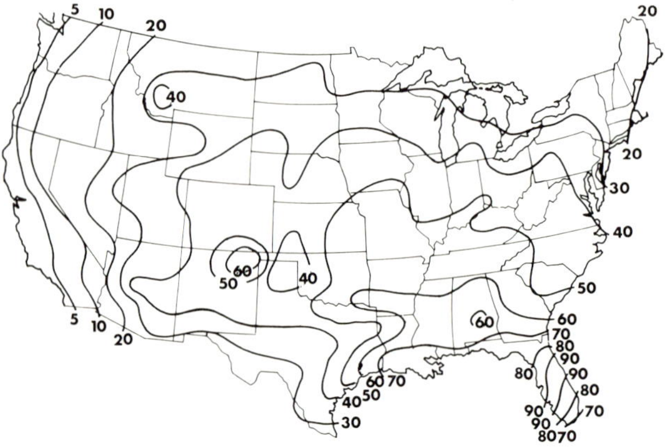

Fig. 46 The distribution of charge within a thundercloud and on the ground. Cloud-to-cloud and cloud-to-ground discharges are shown.

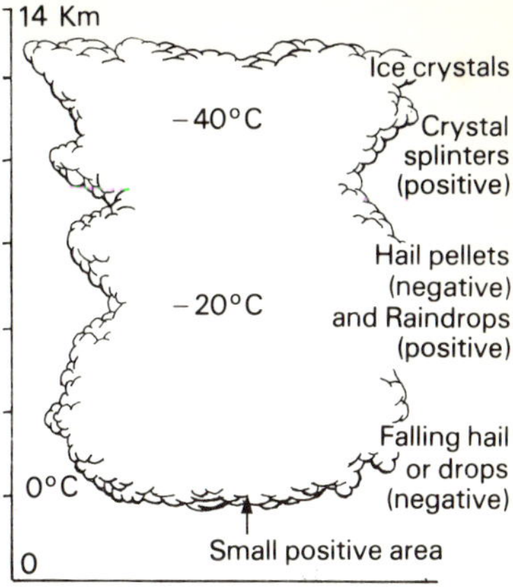

Below Thunderstorms frequently become organised into lines extending across country. A typical anvil shape can be well seen in this photograph of such a line of storms.

Most of the world's thunderstorms are to be found in tropical regions where it is hot and humid. Elsewhere thunderstorms are a distinct possibility when tropical air swings up from the warmer seas. The biggest thunderstorms form in tropical air and grow into the strosphere.

The thunderstorm sequence, or part of it, is also likely to occur when a deep pool of cold air moves down from the polar regions, particularly if its track is over the sea so that it is a polar or arctic maritime air mass. If showers fall, however, these can have the effect of arresting the sequence, and thunderstorms are less likely to develop afterwards. Predicting whether the sequence is actually going to lead to cloud-

Most thunderstorm activity occurs over equatorial regions. Here gleams of sunlight filter through several storm cells over Lake Naivasha in central Kenya.

bursts and falls of hail or not is outside the scope of the amateur forecaster. Professional forecasters or anyone with access to a radar scanner in an aircraft or boat will notice very strong echoes as an active cumulonimbus heavily laden with rain and hail darkens the sky. Cumulonimbus clouds in polar air can usually be seen approaching some 80 km (50 miles) away; they are identified by their distinctive high-level anvil tops, and they cause the wind to become squally when they are imminent.

Individual storms moving up from the tropics are, however, a little more predictable and the conditions preceding a heavy burst are clear cut. The anvil tops are higher than their polar counterparts and can therefore be seen at a greater distance. The whole day feels heavy and sticky as the sky becomes overcast and dark, occasionally glowing white as sheet lightning heralds the approach of the downpour. The heaviest downpours are frequently associated with troughs in the pressure field; but to pick this out with just a single barometer is virtually impossible, and would in any case require a continuous watch to establish the time of most rapid fall of atmospheric pressure.

Thunderstorms are also often associated with the passage of warm and cold fronts and they are concentrated in zones of perhaps 25–80 km (15–50 miles) in width and even greater length, rather than individual storms as found in distinct air masses. Frontal thunderstorms are often difficult to identify as the cumulonimbus clouds are obscured by the lower layer clouds, although pilots of aircraft may find them easier to identify.

Just as ordinary clouds can be initiated or increased in strength by orographic uplift over hills (*p.62*) so thunderstorms may build up over a range of hills or mountains, and at first sight appear to be a form of frontal belt.

It frequently happens that cumulonimbus clouds become organized along a line of instability, often known as a **squall line**, which may extend for hundreds of kilometres or miles. (It should be noted that this term is sometimes used for the roll cloud created by the strong opposing air currents at the base of cumulonimbus.) The intensity of such squall lines produces a high-pressure area (of a few mb) where precipitation is most intense, and a low behind the storm.

The lightning strokes in this photograph, taken with a time exposure, all show the characteristic forked shape. The first branch to reach the ground provides the path for the main ground-to-cloud discharge.

Tornadoes, waterspouts and other whirlwinds

Tornadoes – or 'twisters' – form under the same conditions as severe thunderstorms and squall lines, with which they are very frequently associated. High humidity, extreme instability and marked surface air convergence, that is low and falling pressure, are all required. They are common in the Gulf and Central States of the U.S.A. when warm humid air from the Gulf of Mexico encounters colder polar air, or is overlain by continental tropical air from the south-western states. Such conditions are signs upon which the initial tornado warning service is based, but it would appear that tornadoes – albeit less destructive ones – may form from any intense storm. The sign to watch for is when one of the black bottoms to the clouds starts to sprout a small-radius funnel of cloud down towards the ground. As the up- and down-currents within the parent cloud intensify and subside so the funnel of cloud extends and contracts, sometimes lifting right off the ground for a while before descending again like an elephant lowering its trunk to water.

Tornadoes' extreme destructiveness is a result of their generally small diameter, exceptionally low central pressure – unknown, but certainly well below 900 mb – and extreme wind speeds – again unknown, but in some particularly violent cases probably in excess of 600 kph (over 370 mph). Upcurrents exceeding 240 kph (149 mph) have been recorded and these may lift all forms of objects into the air, quite commonly automobiles, and even on one occasion having sufficient force to heave a locomotive from the track.

Tornadoes frequently occur in swarms, over 100 in a single day have been noted, which may mean that extensive damage is caused to the afflicted districts, despite individual tornadoes being less than 2 km (under 1 mile) across and ranging down to a few hundred metres or yards. These may be quite erratic in their behaviour, some being known to remain stationary for three-quarters of an hour; others following tracks for over 450 km (280 miles), and lasting for 7–8 hours.

Tornadoes die down when the violent upcurrents within the thundery parent cloud subside as the cloud grows past its maturity into the dissipating stage, usually as the sun sinks in the sky. This also applies to **waterspouts** which may be described as tornadoes occurring over the sea, but these generally appear to be rather less intense. Rather different, and certainly less strong, are the various **whirls** or **whirlwinds**, called by all sorts of names, such as **dust whirls, sand devils, water devils**, as well as many local variations. These are produced under conditions of intense local heating and a localized horizontal wind shear, which starts the rotation of the rising column of air. Some of the strongest, such as the land devils formed in south-west Arizona may be as destructive as small tornadoes, but generally their lifetime is much shorter, typically only a few minutes.

A land devil photographed in Kenya, with a smaller example in the distance. The internal structure is well shown by the dust which has been raised by the narrow vortex.

Tropical storms

Among the most destructive weather events in the world are the tropical storms that can produce winds of more than 300 kph (over 180 mph) as well as torrential rain. They play a big part in the climate of the tropics where they form from existing disturbances (although not in the Southern Atlantic Ocean where the seas are too cool). When the tropical air is exceptionally unstable and the convergence zone created by the trade winds of the two hemispheres is displaced over unusually warm areas of sea and is also far enough from the equator for rotational forces to come into play, vast amounts of water are evaporated into the air and carried upwards in a spiral of strong winds. The process is self-reinforcing because the enormous quantity of latent heat released when the vapour condenses into cloud makes the air even more buoyant. Huge towering cumulonimbus grow up to the tropopause, relieving pressure at the surface, encouraging stronger surface convergence and causing intense spiralling winds and strong upper air divergence at the tropopause. Equilibrium seems to be reached when the storm grows to 150–750 km (to about 90–450 miles) across so long as the sea temperature is above 26°C (79°F).

Fig. 47 Isobars around a tropical cyclone (*top*) moving in the direction shown. Areas of precipitation are shown. A sectional representation (*bottom*) shows how air is subsiding into the eye.

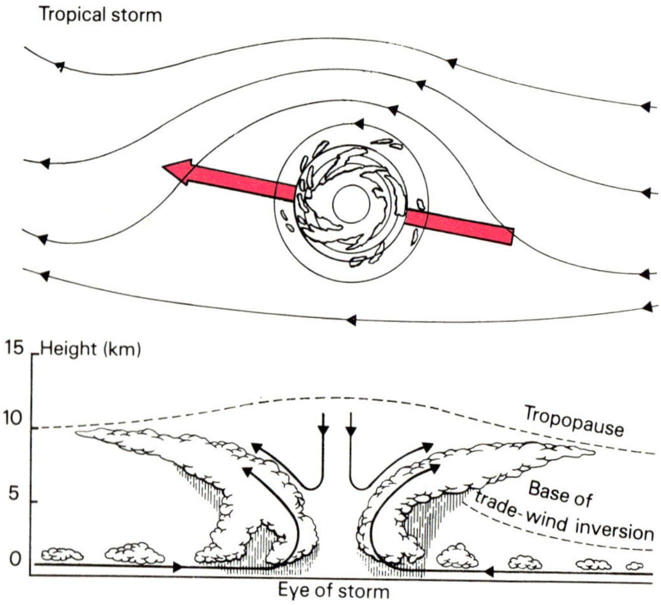

At this stage the storm consists of concentric rings of cumulonimbus-type clouds producing torrential rain and cyclonically circulating winds blowing at some 75–300 kph (about 45–180 mph).

A curious phenomenon is that the centre of the storm – the 'eye' which is some 30–50 km (20–30 miles) across – is almost calm and cloud-free. In fact the air in the storm's eye is warmer than the surrounding air by some 3°C (5°F) and is actually sinking. Why this should be so is not certain, but the most likely explanation is that the complex interactions and dynamics of the storm bring about the sinking in order to replace the air exported at a high level – where a wide cirrus shield is formed – and that the latent heat released contributes to the warming. Pressure drops rapidly towards the centre and may become very low indeed – the lowest known being about 877 mb. If and when the storm tracks inland the increased friction deflects the circulating winds towards the eye so that it dies down, and in any case removal of the source of latent heat – the warm seas – eventually kills the storm. Over cooler seas an intense depression frequently remains and its origin may sometimes be recognisable by the generally circular pattern of its isobars. Such systems may remain distinct and travel long distances; one such hurricane started near

Fig. 48 Typical tracks of tropical cyclones, shown in red (May-November) and black (November-May).

This Apollo 7 photograph shows the eye of Typhoon Gloria south-east of Okinawa. The eye's diameter is about 80 km (50 miles) and the tops of some of the convective towers can be seen above the main cloud sheet.

Fig. 49 A chart of the sea areas around the U.S.A. showing the hurricane warning offices and the regions for which they are responsible.

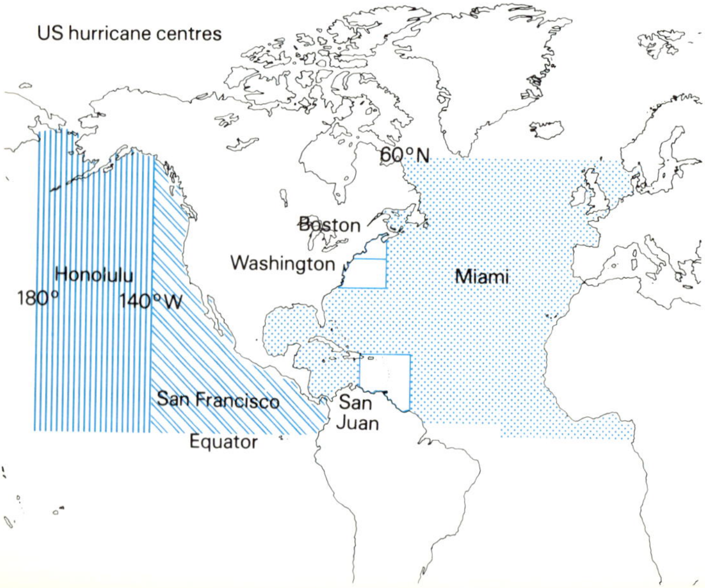

Cape Verde, crossed the Atlantic, turned parallel to the eastern American seaboard, and disappeared near the North Pole.

Over the West Indies, the Gulf of Mexico and the eastern seaboard of the U.S.A., tropical storms are known as **hurricanes**. Over the West Indies the easterly trade winds prevail, and occasionally during the period from August to October hurricanes devastate the islands. The hurricanes, like all tropical storms, feed on the warm tropical seas to remain self-perpetuating, and so the storm centre often turns on to the warm Gulf Stream and follows a path near to the East Coast. Even though the eye of the storm may be out at sea the American coastal regions can be ravaged by the periphery of the storm. The amazingly strong winds flatten well-nigh everything in their path.

Tropical storms are equally destructive wherever they are found, but names vary in different parts of the world. They are **cyclones** over the Bay of Bengal and the Arabian Sea; **typhoons** over the China Sea; **willy-willies** over Australia. Off the east coast of Southern Africa they are called by the true name for the whole class **tropical cyclones**.

North American weather

Much of North American weather is affected by the significant barrier to the westerly winds that is formed by the Rocky Mountains and the subsidiary ranges. These mountains receive most of the moisture from the maritime tropical air off the Pacific, giving rise to the much drier conditions over the Great Plains. Westerlies crossing the mountains can give rise to the famous Chinook (the 'snow-eater'), the föhn wind on the leeward side which has been known to raise temperatures by 22°C (40°F) in 5 minutes. Many depressions weaken over the ranges, but reform on the east, become very strong, and move south-eastward then north-east, greatly influencing the weather over the Great Plains and the eastern states. Off the West Coast, warm air streaming over cold upwelling water frequently gives rise to extensive sea fog, particularly in summer.

Arctic air (built up over the Canadian Arctic islands) and continental polar air (from northern Canada) often sweep down through the central and eastern regions to bring cold air as far as the Gulf States. This air stream is frequently helped by the strong, cold highs over the prairies in winter. Arctic Front depressions frequently move south-east, merging with others produced at the Polar Front to give severe blizzards over the Great Plains, the Great Lakes and the eastern seaboard. The Great Lakes themselves contribute considerable moisture to the air flow in early winter, leading to especially heavy falls to the south and east.

In the south the way is open for maritime tropical air to penetrate in summer up the central valley of the Mississippi and beyond into Canada. The instability of this air is largely responsible for the high incidence of violent thunderstorms and tornadoes in the central states. This is accentuated when continental tropical air flows from its source region over Arizona and the other south-western states, above moist air from the Gulf of Mexico.

Depressions generated at the Pacific Polar Front frequently decay after the long ocean crossing before they reach the land. This contrasts with the situation on the east, where the Polar Front may reach down as low as the Gulf of Mexico. Depressions generated at low latitudes grow as they travel up the eastern coast, feeding on the temperature difference between cold continental air over land and warm maritime air over the Gulf Stream, and being steered in their course by the Appalachians, while depressions formed over Colorado follow a parallel track further to the west. A somewhat similar path is taken by the destructive hurricanes, generated over the heated seas of the southern North Atlantic. Unlike the West Coast, the Atlantic seaboard has more of a continental than a maritime climate. Just as on the west, though, warm moist air passing over the cold Labrador Current gives rise to extensive fog over the Grand Banks off Newfoundland.

Fig. 50a, b The main motions of air masses over the North American continent **a** and the general climatic zones **b**. Also shown are the average monthly temperatures and rainfall for the reporting stations marked.

Annual rainfall in millimetres

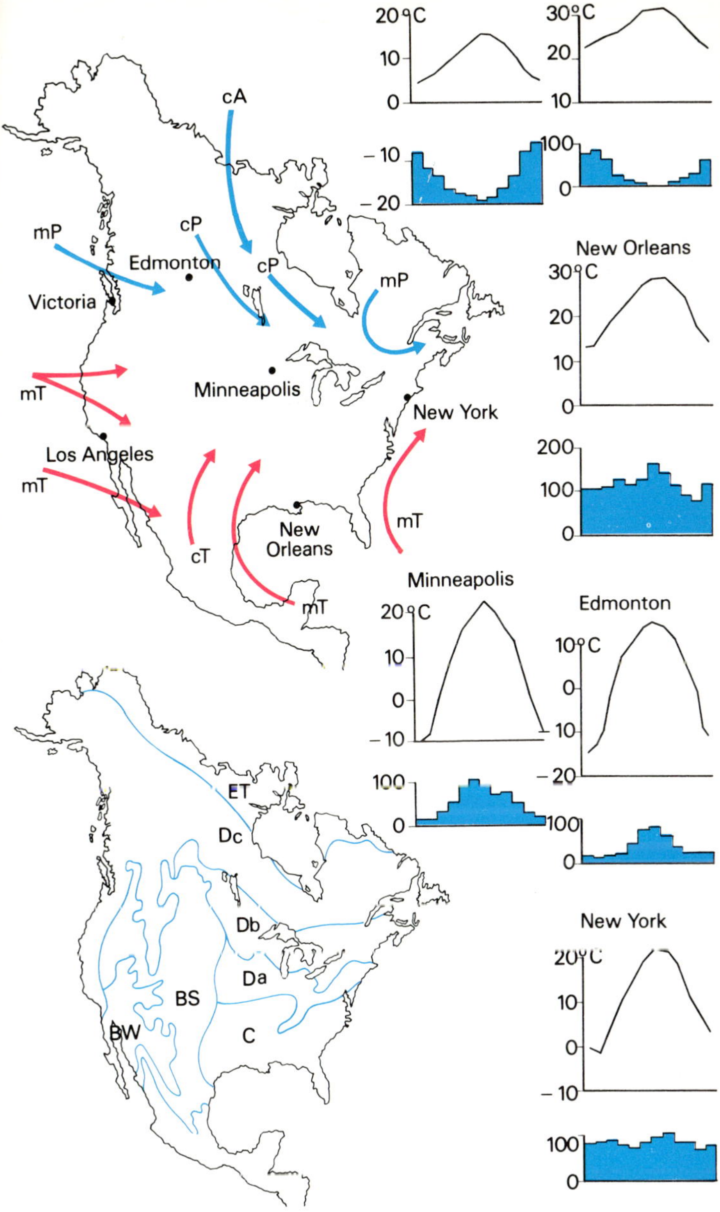

European and Mediterranean weather

The weather over Europe, especially Western Europe, is notoriously variable, since it is affected by nearly every type of air mass, although the east has a more truly continental-type climate. There is a very high frequency of low pressure systems which travel in from the warm Atlantic and may penetrate far into the continent, especially in early winter. However blocking anticyclones are fairly frequent, particularly over Scandinavia, and these divert the depressions northward over Iceland or south-east and north of the Alps. The frequency of the blocking highs is very variable, so that they may sometimes bring continental polar (or even arctic) air and exceptionally severe weather to Western Europe, and at other times they bring either fine, warm or hot weather, or seemingly endless, dull days when depressions linger over the area. High pressure patterns can also bring waves of the cold, but fairly dry, maritime polar air.

Warm, moist maritime tropical air from the Atlantic encounters cold sea surfaces off the western coasts producing extensive sea fog, although the warmth of the sea is sufficient to make the general climate significantly milder than might otherwise be expected for its latitude. Maritime tropical air usually forms the warm sector of most depressions. When it encounters cold polar or arctic continental air from the east, considerable snowfall can occur.

The Scandinavian mountains produce significant differences in the climate on the two sides of the range, and both they, and the Alps, have considerable steering effects upon the paths of depressions. In the second case low pressure to the north can draw warm Mediterranean air over the mountains to give föhn conditions on the northern side.

The Mediterranean itself has a distinct climatic regime, to which its name has been given. This is characterised by considerable winter rainfall followed by hot, dry summers. It is dominated in summer by the sub-tropical high pressure region, and continental tropical air from Africa. In winter the sea's warmth, a generally westerly air stream and an additional front (the Mediterranean Front) give rise to a succession of depressions which travel eastwards along its length. Such depressions, travelling east, are very frequently regenerated at the eastern end of the Sea by continental polar air from western Russia, and themselves tend to swing north-east over the area of the Black Sea. The complicated relief on the northern side gives rise to considerable modification by various local effects. The mountains too, are responsible for the strong katabatic (*p. 68*) winds of the Mistral and Bora (the latter affects the head of the Adriatic). From the African side, the Sirocco and Khamsin winds bring hot, continental tropical air northwards, where it may pick up considerable moisture from the sea.

Fig. 51 Predominant movements of air masses over Europe and the Mediterranean **a**. The temperature and rainfall diagrams illustrate the difference between the maritime, continental and Mediterranean climatic zones **b**.

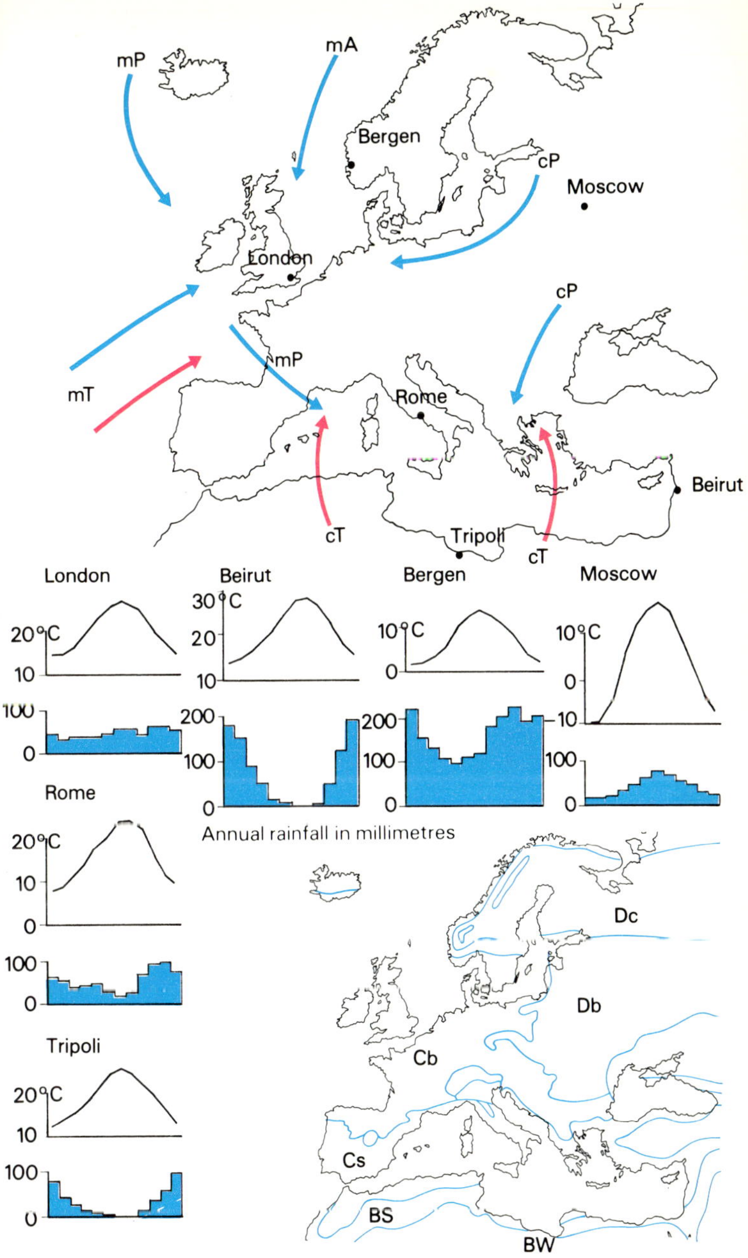
mP
mA
Bergen
cP
Moscow
London
cP
mT
mP
Rome
Beirut
cT
Tripoli
cT
London
Beirut
Bergen
Moscow
20°C
10
30°C
20
10
10°C
0
10°C
0
-10
100
0
200
100
0
Rome
Tripoli
Annual rainfall in millimetres
Dc
Db
Cb
Cs
BS
BW

The weather of Australia and New Zealand

Australia is sufficiently far north for it to come well within the influence of equatorial air, and indeed the north-western coast in particular is subject to a monsoon regime with north-westerly winds in summer and south-easterlies in winter. Hot and humid maritime equatorial and maritime tropical air readily penetrate over the central desert regions as far as 21°S, and may bring rains even farther south. However the mountains of the Great Dividing Range in the east runs along the complete length of the coast and prevents easterly maritime tropical air from penetrating very far inland. The effects of the Range are seen in the maritime climate found in the coastal strip, while on the western side the climate is much drier and of a more markedly continental type.

The interior itself is a source region for continental tropical air, and the air mass which builds up is warm and dry in winter. In summer the intense heat produces a very hot, dry air mass which in the form of a very hot, dry and dusty wind is known as the Brickfielder.

In the south, maritime polar air is the predominant influence, with a constant series of depressions in winter bringing rain to coasts in the south. In summer, the general global southward migration of the dominant pressure areas causes these depressions to follow tracks at higher latitudes, so that only Tasmania is occasionally affected. These southern regions then, have a Mediterranean type of climate, with significant amounts of rainfall in winter, but dry, warm summers. Sudden changes of wind direction from northerly to southerly with incursions of very cold air affect the south and south-east.

Destructive tropical cyclones can affect the north of the continent especially the areas close to the Arafura and Timor Seas. It is not unknown for these storms to follow long tracks right round the western coast of the continent. The North Island of New Zealand, too, may feel the effects of similar storms, originating in the western Pacific.

Being far from any major land mass, New Zealand does not experience the influence of continental air, and the dominant types are maritime tropical and maritime polar. These are present in the endless series of low pressure systems which cross the country with the prevailing westerly winds, the Roaring Forties, encircling the globe between latitudes 40°S and 50°S. The relief, particularly of South Island, leads to heavier rainfall on the west coast, and greater sunshine in the east, sheltered from the north-westerly winds. In the absence of continental influences there are only rare periods of settled weather, but the short distance to the Antarctic does mean that the maritime polar air is colder than comparable air in the northern hemisphere.

Fig. 52 Air masses **a** and climate **b** of Australia and New Zealand. Average temperature and rainfall diagrams show conditions at the reporting stations indicated.

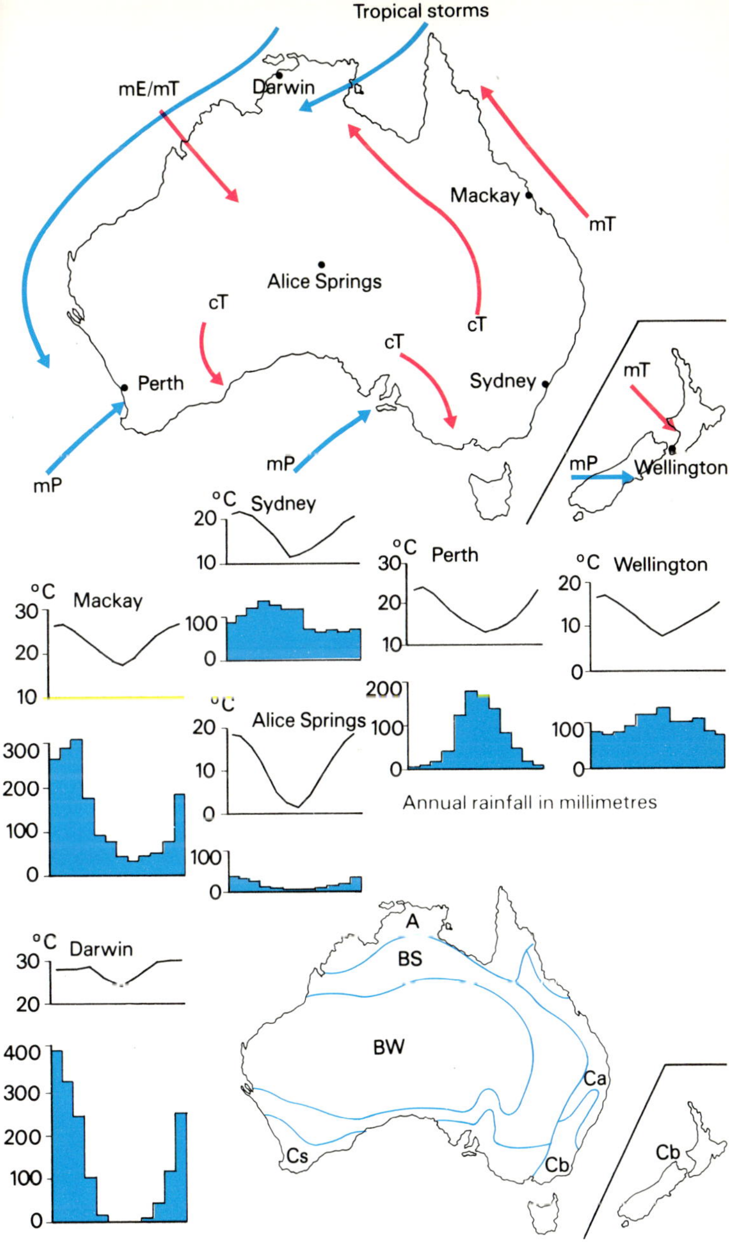
Tropical storms
mE/mT
Darwin
Mackay
mT
Alice Springs
cT
cT
cT
Perth
Sydney
mP
mP
mT
mP
Wellington
°C Sydney
20
10
100
0
°C Perth
30
20
10
200
100
0
°C Wellington
20
10
0
100
0
°C Mackay
30
20
10
300
200
100
0
°C Alice Springs
20
10
0
100
0
Annual rainfall in millimetres
°C Darwin
30
20
400
300
200
100
0
A
BS
BW
Ca
Cs
Cb
Cb

Professional forecasting

A study of the foregoing sections should stand you in good stead to try your hand at forecasting your own weather – and, like everything else, the more attempts you make the more accurate you will become. But there is a great deal to be gained by comparing notes with the professional weather forecasters. Official weather forecasts are based on simultaneous weather observations from all over the world. There is an extensive international network for the exchange of the necessary data, with a 'Main Trunk' route running from Washington, D.C., in the U.S.A., through Bracknell in the United Kingdom, Offenbach in West Germany and on to Melbourne in Australia. All the main national weather centres are linked into the primary route.

The National Weather Service in the U.S.A. and the Meteorological Office in Britain, for example, are primarily concerned with forecasting American and British weather respectively, and employ some of the largest and most powerful computers in the world to digest the enormous input of data that streams in every hour of every day and every night from all around the Northern Hemisphere. The computers process the data, working under a set of rules which best represent our understanding of the physics of the atmosphere. The speed at which a well-programmed computer can work is so very fast that the current hourly trends can be projected rapidly into the future (subject to certain limits and conditions). In this way forecasts of the broad weather patterns for up to five days ahead may be produced. A team of forecasters then apply their skill, experience and judgement to the computer readout. They try to identify major errors and check for consistency at all levels of the atmosphere. They can even check the computer forecasts from different countries against each other. Since no computer model of the atmosphere has a small enough grid of input observations in time and space, the forecast charts it produces will not be precise. In the short term (1–6 hours) local land influences that cannot be included in the models cause errors. In the longer term (two days and more) small initial errors snowball. Britain in particular also suffers from the scarcity of observations made out in the Atlantic, which is the area from which its weather typically arrives. Other maritime areas, such as New Zealand, suffer from similar problems, which is where the local weather stations come in. The forecasters there are fully informed of the computer-based forecasts for the broad-scale weather patterns. Using their experience and knowledge of how the current weather is shaping they are able to update their charts and act as interpreters between the computers and the public.

Radar can be used to monitor rainfall continuously. Yellow shows the area of heaviest rain in this belt over Wales, light and dark blue indicating moderate and light amounts.

Fairly conventional radar units give considerable information about the distribution of heavy rainfall as shown on this airport display. Vertically scanning radar can give a picture of the distribution within individual showers. For further information about the use of radar see *p. 130*.

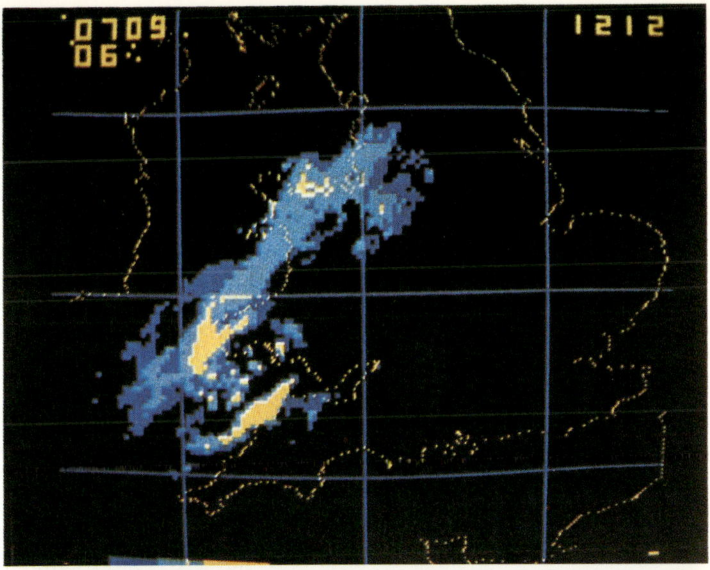
0709.
06
1212

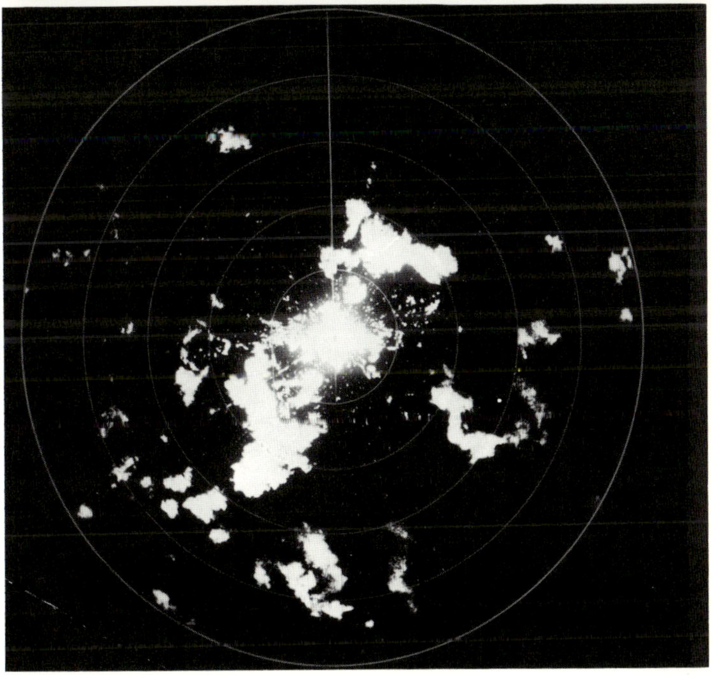

Understanding forecasting language

Forecasting the weather itself is relatively easy. Forecasting the exact timing, though delightful when it comes off, is really very difficult. Because of this difficulty weathermen have developed their own jargon to express the inherent uncertainty in every forecast. Ask a forecaster, 'Will it rain tomorrow?' and if he answers 'No', ask again, 'Are you really sure?' and nine times out of ten the answer will be, 'Well, there is just an outside chance that if . . .' or something similar. There is always more than one road the weather can go down. The forecaster selects the most probable.

There will be times, however, when a person who is unfamiliar with the terms will be totally confused by forecasters' brave attempts to cover all the possibilities. To some extent the use of jargon has been forced upon the weathermen by the need to condense the forecast as much as possible. Since it is not possible to forecast accurately the exact coming and going of cloud and sunshine hours (or clear sky hours at night) recourse is often made to three terms: 'intervals', 'periods' and 'spells'. Although not specific, all nevertheless imply some sunshine (or clear sky) hours in respectively increasing amounts. A day of sunny spells has more than half the daylight hours filled with direct sunshine. A dry spell consists of at least two days without precipitation.

The number and frequency of showers is something else that cannot be accurately forecast. So showers are described as 'isolated', 'scattered' or 'frequent'. Isolated showers can generally be regarded like burglaries, as things that happen to other people. Here the dominant factor is whether or not some local topographical feature affects the airflow enough to produce a localized shower. Away from shower-prone localities scattered showers are very few and far between, to the point of not turning up at all on many occasions when forecast. Frequent showers imply that almost everyone will see at least one. The use of phrases such as 'rain here and there', 'at times', 'on and off' simply reflect the fact that rain seldom persists throughout a 24-hour period, but often moves in cells of rain along the line of rain-bearing clouds. Exact timing of the arrival of the rain cells over specific locations is therefore impossible.

It is also necessary when watching, listening to or reading a weather forecast to work out the degrees of probability of any forecast event. In North America the probabilities of a certain weather event happening are stated all the time; while in Britain and elsewhere the probability is sometimes implied by the choice of words. 'Risk of', 'perhaps', 'likely', 'expected', 'imminent', are increasing degrees of certainty that the weather event will turn up. 'Rain' without qualification is as near to being a certainty as a weather forecast ever gets.

Television forecasts and isobaric charts

A forecast that has been overtaken by events is of no use to anyone. Therefore rapid dissemination of the forecast to the public is essential. This is best achieved by television, which allows the forecast to be accompanied by illustrations and where the information given can be right up to date. The best time to watch is in the early morning when the forecast for the day ahead should be most reliable. Traditionally though most people watch the evening weather reports which cover the period overnight and the following day.

Since the television forecast is a short-lived broadcast crammed with information it is worth considering exactly what is going on in order to be able to separate the wood from the trees – or the 'intervals' from the 'periods'. Although the style may vary from presenter to presenter, the format is always chronological. The present state of the weather and its development over the past 24 hours is first of all described. This may be done with either an isobaric chart, a satellite, image, or all three.

Isobaric charts are traditionally the weather forecaster's most useful tools and to anyone who has learned to interpret them, isobaric charts of present and future weather are considered essential. Unfortunately, most people still find a glaze comes over their eyes when confronted by a map full of lines and circles. But there is nothing at all difficult or mysterious about isobaric maps. If the barometers at, say, four different land stations all point to the same figure (the same atmospheric pressure) then you would be justified in drawing a line, an isobar, connecting all four places. Why you should want to do this will become clear later. If you were then given half a dozen or so different barometer readings from ships or meteorological buoys at various positions out in the ocean, you could soon figure out roughly on which side of each reporting point the isobar is and where that same isobar should go if you wanted to continue it out across the sea. Then choosing, one at a time, other atmospheric pressure values,

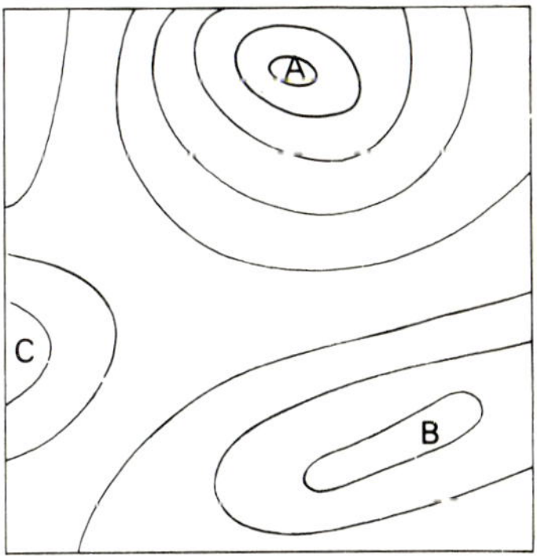

Fig. 53 The dominant pressure system at any place is easily seen from an isobaric chart. At A, low pressure dominates and high at B, while C is under a weak ridge of high pressure.

all separated by a common interval, the exercise can be repeated, adding another line until finally the next highest pressure value and next lowest pressure value do not appear anywhere on the map. Therefore straightaway the areas of highest and lowest pressure will be evident. The areas of settled and unsettled weather can be spotted at a glance by simply picking out the 'highs' and 'lows'. Also remember that as explained (*p. 18*), winds above 500 metres (1 640 ft) or so (low cloud level) blow along the isobars in accordance with Buys-Ballot's law, with low pressure on the left in the northern hemisphere. In addition surface winds, retarded by friction, blow about 20° across the isobars from higher to lower pressure.

So when you look at an isobaric chart note how near the nearest high and the nearest low are and try to decide which is the dominant feature and which way the wind is blowing. How do you decide this? If the isobars that are running across the local area appear to be curving round the parent high centre then the anticyclone is dominating the weather. If on the other hand the isobars appear to be curving round the parent low pressure centre then the depression is dominating the weather. This is because high pressure centres spread out their influence in the form of **ridges**, causing the isobars to curve towards them. Low pressure centres extend their influence in the form of **troughs**, also causing the isobars to curve towards them. Ridges and troughs should next be identified and thought of as belts of weather which are becoming, respectively, settled and unsettled. When the curve of the isobars that identify a trough become so sharp that they become like the letter *V* (lower pressure to the top), then the trough is known as a front. It is a zone of maximum surface air convergence (*p. 9*), and therefore uplift, and therefore cloud, and most likely precipitation. So fronts should next be identified and thought of simply as zones of bad weather. The sharpness of the isobars at a front often gives a good indication of its severity, since the sharper the isobars the stronger the low level convergence. (Actually, the severity of the front would be better determined by the temperature

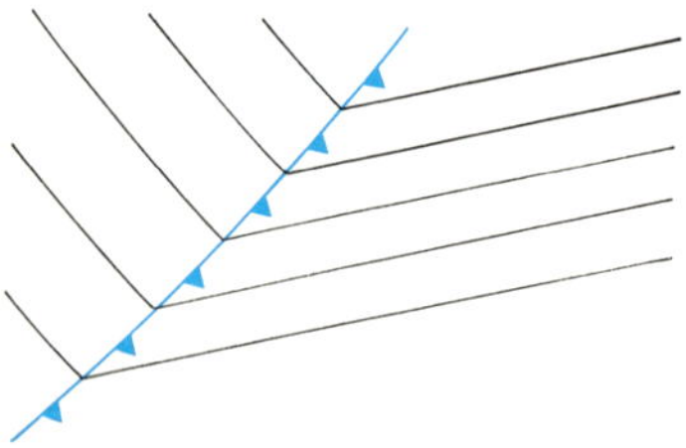

Fig. 54 The location of a front (in this case cold) is shown by the sharp change in the direction of the isobars, giving a *V* shape.

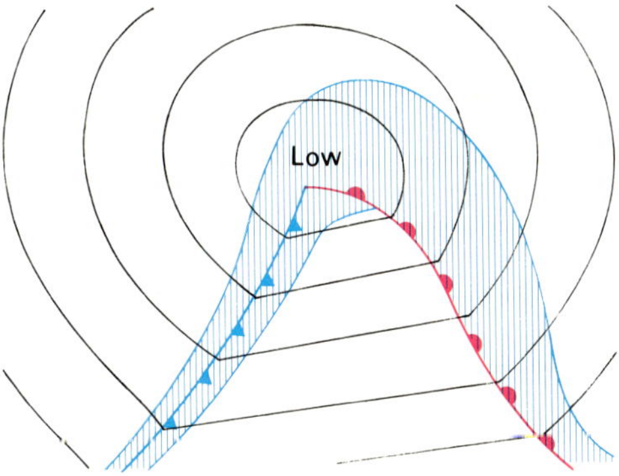

Fig. 55 A schematic diagram showing the zones of maximum rainfall in a depression system. Apart from this, drizzle and fog are likely in the warm sector and occasionally showers, which may also occur behind the cold front.

change across the front, but televized isobaric maps do not usually show temperatures.)

Next, notice if the surface fronts are warm or cold. Warm fronts herald the approach of warm air and slope upwards in the direction of travel so that precipitation arrives *before* the front. Cold fronts announce the approach of cold air and slope backwards, so that precipitation arrives *with and after* the front. Clearly fronts which appear to be drawn as shallow troughs in an otherwise broad ridge will be weak affairs, probably amounting to no more than an increase in cloud.

Overriding all these points is the accuracy of the observations made – it is as well to remember that the isobaric chart is only as good as the observations from which it was drawn. Since the number of observations is always small in relation to the area that the chart covers, it is by no means certain that the positions of the highs, lows, troughs, ridges, and fronts are all fully correct, or even that some have not been missed out completely. Although not usually shown on charts issued for general public use, detailed information about conditions at each reporting station are plotted by the use of standard symbols.

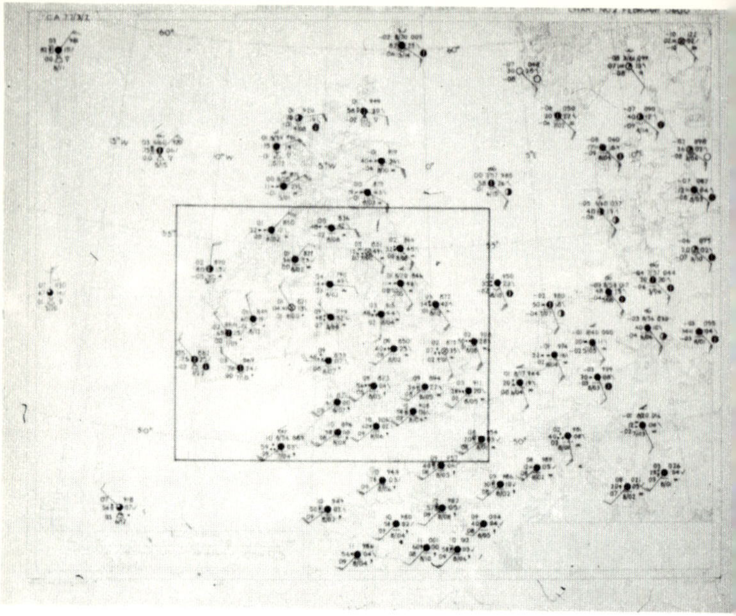

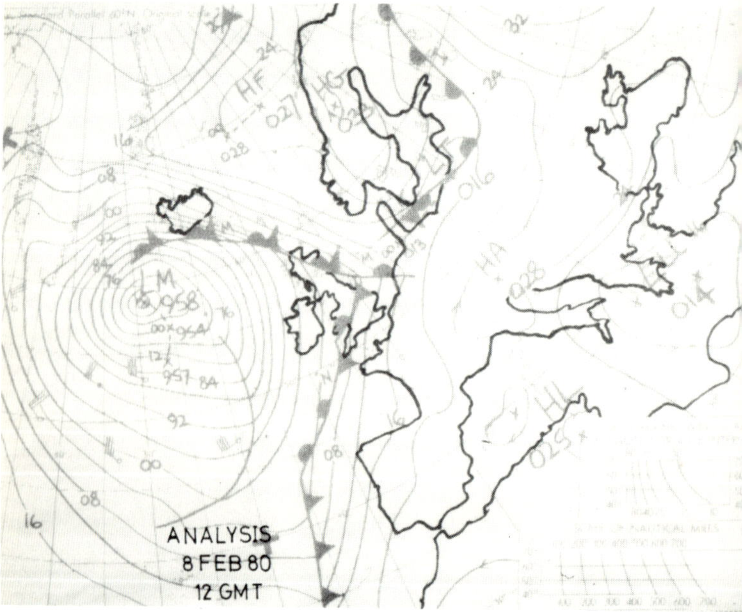
ANALYSIS
8 FEB 80
12 GMT

Above A plot showing the computer-predicted pressure pattern for a large part of the northern hemisphere. Such predictions provide a basis for more detailed forecasting of actual weather developments.

Opposite top A computer-plotted surface chart for north-west Europe. The standard symbols are used for data from all the reporting stations.

Opposite below A surface analysis chart for the North Atlantic, Europe and the Mediterranean, where isobars and fronts are shown. Note the secondary depression developing off the Bay of Biscay.

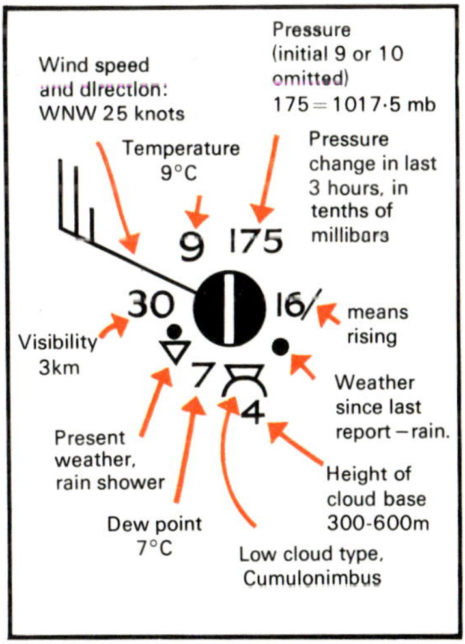

Fig. 56 The way in which station data are represented on synoptic charts.

Interpreting satellite photographs

A representation of the current weather pattern which is undeniably correct is the satellite image. There can now no longer be any argument about where there is cloud and where there is none. The way forward now in the field of professional short-term weather forecasting is in combining the satellite and radar data as a sequence of time-lapse pictures in colour on the television screen. There are two types of satellite and two types of picture used by weather forecasters. One type of satellite orbits the earth from pole to pole, scanning the atmosphere and ground directly below as it goes. During each successive orbit (of 100 minutes) the Earth has rotated slightly (25° longitude), so it photographs a different pole-to-pole strip at a later time. The other type of satellite orbits the earth over the equator at exactly the same rate as the Earth rotates, so that it is always over

An infrared image of the whole Earth, obtained by a geostationary satellite. Coldest areas are the lightest in tone, NW Africa being the warmest region shown here.

exactly the same position – it is **geostationary** – and scans the whole Earth disc every 30 minutes from 22 000 miles out in space. In this way the movements of the clouds can be watched. When the satellite is in the Earth's shadow at night there is no visible light reflected from the Earth's disc, but the invisible terrestrial heat radiation – the infrared – is still there. The thermal map of the Earth's disc can be produced every 30 minutes day and night. The brightest radiance represent the deepest, coldest clouds where the precipitation probability is about 90%. The darkest radiance represents the hot spots of the warmest land. The shades of grey between represent contrasting land and sea temperatures and clouds at different altitudes. The visible (daytime) image consists of the same full range of tones between black and white, but in this case the image produced is more like a normal black and white photograph – the most reflective surfaces such as cloud tops and snow are displayed as the brightest tone and the least

A very deep depression photographed over the Atlantic by NOAA-5 polar orbiting satellite. The long occluded front can be distinguished as well as the wide belt of the warm front and the narrower cold front running towards the bottom.

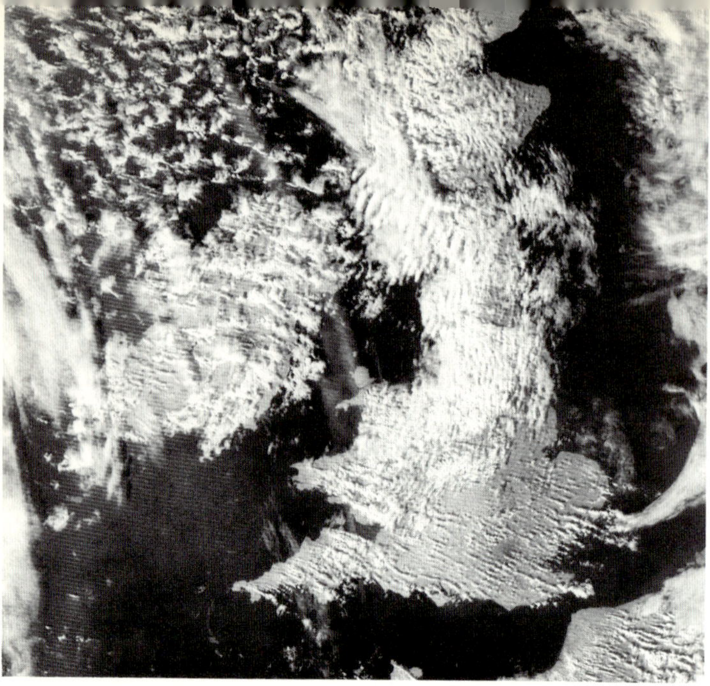

Above Organized cloud over the British Isles. Wave structure can be seen in the medium and high clouds over Scotland and northern England. Long streets of convective cumulus run across Wales and southern England. The wind was from the west.

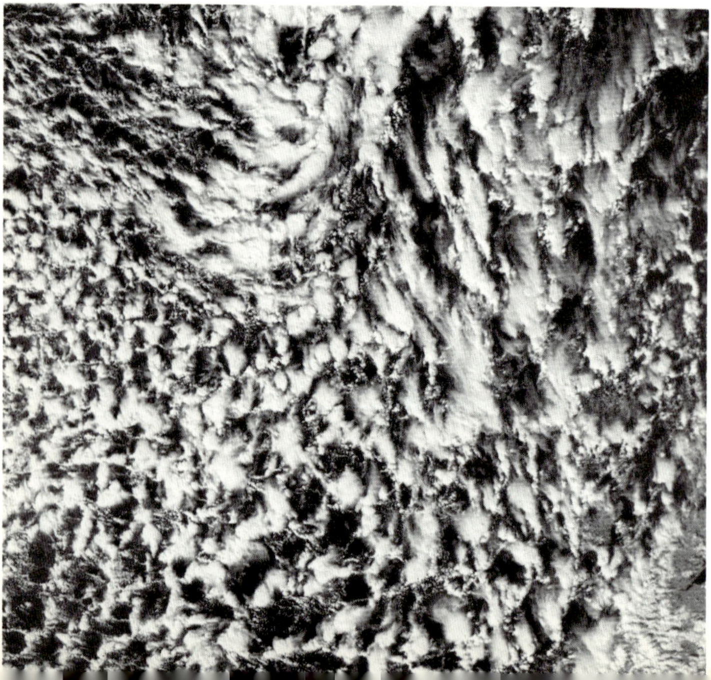

reflective surfaces such as forests and dark soils as black, with all the shades of grey in between.

So when looking at a satellite image you first of all need to know whether it shows the infrared part of the spectrum or reflected visible light. If the edge of the Earth's disc is shown remember that outer space is both cold (so that on an infrared image it appears white) and dark (so that on visible light pictures it appears black. Clouds appear white or light grey on both sorts of picture, so the absence of these tones means that the air is cloud-free. Remember also that fog does not show up on infrared images because it is at almost the same temperature as the underlying surface.

Next notice the shape and extent of the clouds. Depressions can be picked out as swirling spirals of cloud. Long, fairly organized and definite belts of cloud are usually cold fronts, and the fat bulge of cloud ahead is usually the warm sector and the warm frontal cloud. Individual speckles of cloud appearing at random typically indicate fair weather cumulus in polar airflows. In winter the speckles are mostly over the sea, whereas in summer they appear mostly over the warmer land. A honeycomb of white 'cells' indicates that the air is highly unstable and that showers are probable. Organised lines of cellular cloud radiating from somewhere near a depression centre are zones of showers or troughs, and the brighter the cells the more likely it is that the showers will be thundery.

A further good indication of storm-like showers is the presence of comma-shaped cells of cloud, and also clusters of cells merging together. Another type of area of heavy rainfall can be identified on a long sharp belt of cloud, where a slight hump breaks the otherwise smooth sweep of the trailing edge of the cloud.

Cloudless areas clearly indicate fine settled weather and are commonly areas of slowly sinking air. Clear areas of sky often follow bands of continuous cloud, since sinking air is necessary to compensate for the zones of convergence and rising air.

Shapeless wedges of not particularly bright cloud which cover relatively large areas are usually sheets of stratocumulus couds that are tropical in origin.

The next step in interpretation is to note the time to which the image relates, and mentally to move the clouds to account for the delay. Remember that clouds move with the wind and can develop or dissolve as they go, so that the cloud seen over any given area may not still be there at the time of the broadcast. However, the satellite image shows clearly the type of airflow moving towards your local area; from this you can work out its potential for deep cloud and therefore for rain or clear skies.

Opposite Cellular structure to the clouds over the NE Atlantic. Note how the cells and their cirrus tops are more organized towards the north, and also how the nature of the clouds changes over Scotland (*bottom.right*).

Radar imagery

Radar images of precipitation rates over a local area are sometimes shown. Heavy, moderate or light precipitation rates, depending on the strength of the radar echo from the clouds, are ascribed different colours. The data from a network of radar scanners can be combined to cover a larger area and the composite image updated every 15 minutes. A time lapse sequence of such images shows the motion and development of the precipitation cells. This technique of anticipating imminent precipitation is called 'nowcasting'. Meteorologists are now working on ways of presenting the isobaric map, the thermal map, and the rate of precipitation map simultaneously, clearly and reliably.

Visible (*left*) and infrared images of NW Europe. A weak low is centred over Scotland, the cold front extending south over France and the warm towards Norway. Note the thunderstorms over the Alps and how the high pressure area centred south of Leningrad is generally clear of cloud.

Radio forecasts

Radio forecasts are usually of two types, the first type is intended for the general public and may be prepared and broadcast on either a national or local basis. These are usually very similar to the television broadcasts, although naturally without the isobaric charts or satellite imagery. The second type is very much more detailed and is very often primarily intended for the use of sailors. (Aircraft pilots will usually be able to obtain the even more detailed three-dimensional information which they require from a local airport forecast office.) This second form of broadcast is usually transmitted at a few specific times during the day and on a limited number of frequencies.

Although practice varies from country to country, it is usual for both land and sea areas to be defined and named, and for outline charts of these to be available. In announcing the forecasts, the details for each of these areas are given in a set, logical order. Similarly actual observations from selected reporting stations are often given, and again the order will remain the same from broadcast to broadcast. From the details given anyone may plot a regional chart with considerable detail. The forecast usually gives a general summary and then gives expected wind direction and force, precipitation and visibility. Actual reports will include these items and also pressure relating to a stated time. The summary and the actual reports will show the positions of highs and lows, and of the isobars. The forecast will give a good idea of the movement and development of the systems.

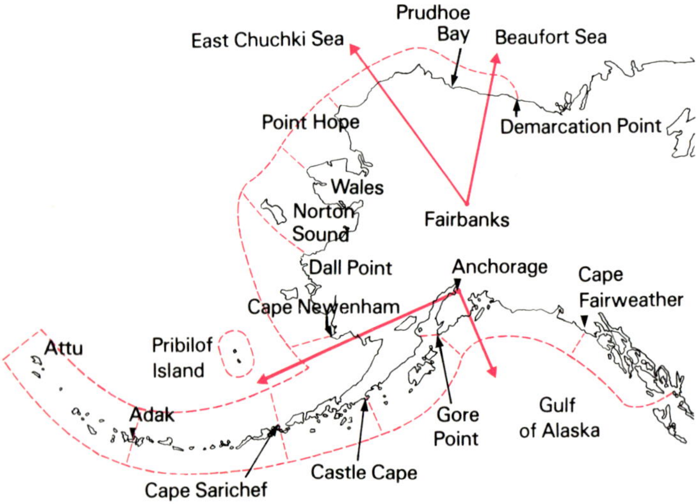

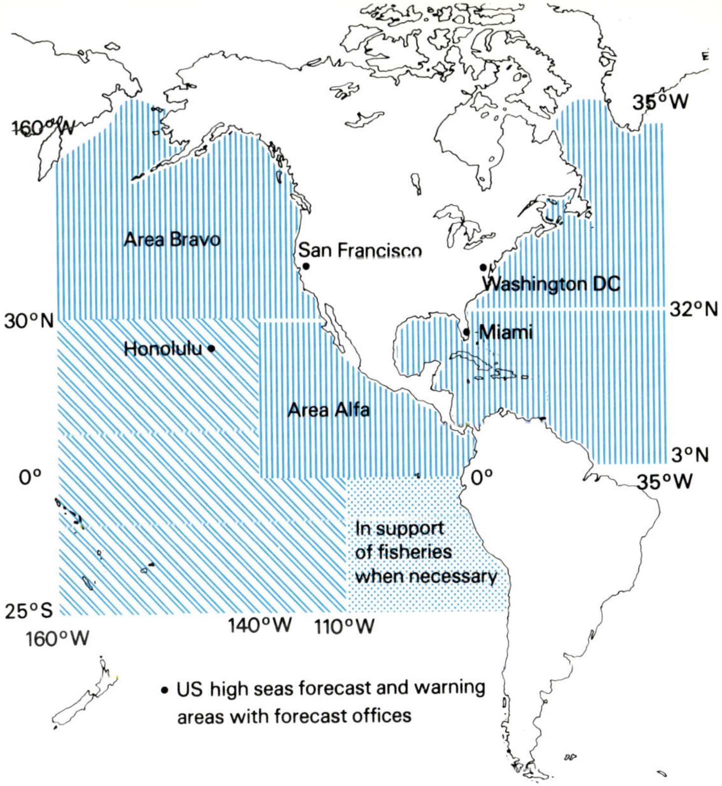

Fig. 58 Most maritime nations prepare forecasts for oceanic areas. Here the U.S. areas are shown, together with the individual offices issuing radio forecasts.

Fig. 57 *left* Typical coastal forecast areas extend only a short distance offshore. Those for Alaska are shown, but NOAA weather radio stations cover the whole of the seaboard of the U.S.A., including the Great Lakes.

Telephone and newspaper forecasts

One of the faults of the television weather forecast is that in the end it is just a memory. A lot of information has to be condensed into a broadcast not lasting longer than a minute or two, and it is inevitable that the importance of certain points will be overlooked in the welter of facts. A similar problem affects radio broadcasts, but here it is more common for listeners to tape record them so that the very detailed information, given in the specialized broadcasts, can be checked again.

A weather forecast that can be listened to time and time again until everything is crystal clear is the recorded telephone weather forecast service, which deserves to be better known. True there are no illustrations and it is rather impersonal, but to the informed amateur all the salient information is there. The script of the tape, prepared by the local weather centre and covering the local area's weather for the next twenty-four hours, is revised about every six hours. Since the time of issue is stated at the start of the recorded message it does not take long to work out exactly when a new tape will replace the old, which is of course the best time to phone in and check the forecast.

Newspaper forecasts have the advantage of being available to be studied at leisure, but they are based on relatively old data. The morning newspapers take their copy from a tightly worded forecast issued during the afternoon of the day before. The forecast isobaric chart for midday is also out of date (but not necessarily wrong) by the time it reaches the reader. The sunshine temperature tables are of course only a guide to yesterday's weather – and who wants yesterday's news?

Depending upon their exact press times, local evening newspapers may use later information from the local weather station.

Apart from isobaric charts with varying amounts of detail, a very few newspapers regularly or occasionally show a satellite image, and although these are usually older than the corresponding TV ones, they can, of course, give useful guidance to weather patterns.

Making use of forecasts

Any form of forecast will usually be divided into night-time and daytime predictions. The essential points to pick out are the expected maximum temperature in summer, and the expected maximum and minimum temperatures in winter. The forecast minimum night temperature should be mentally amended by making suitable adjustments to allow for the effects of local features upon the weather. The weatherman does not have enough broadcast time to consider every place in isolation. So when the general temperature is expected to fall below 4°C (39°F) and you are concerned about the effects of frost, it is worth checking for yourself late in the evening to see at what rate the temperature is dropping and forecasting your own local night minimum in the manner described (*p. 150*).

It is also a good idea to have a clear picture of exactly how the morning is expected to start. When the official forecast goes sadly wrong it is often the morning sky that provides the first clues as to whether the expected weather system has speeded up or slowed down, or a new trend altogether has started. The morning radio weather forecast will confirm or refute your reservations.

Television, radio and telephone weather forecasts are all prepared by meteorologists using nationwide data which is at least two hours old. So on some rare occasions, despite the weatherman's claim that it is dry, it can actually be raining outside. However at some centres (particularly at airfields) there are facilities for scanning the local sky with radar, and they are therefore able to pick up storm clouds up to about 250 km (155 miles) away as they appear. Drizzle cannot be picked up by radar since the reflected echoes are too weak to appear on the monitor screen.

On occasions when the following day's weather is particularly important (because of some special event, such as a wedding) the way to maximize the usefulness of the media forecasts is to note the television forecast during the evening and in the morning check it against the morning sky. If all is not well, listen to the morning radio or television forecast to determine what has changed during the night, and if the story is still not clear phone the nearest telephone-listed weather information centre.

The best way of finding out the weather prospects further afield is to call the nearest weather centre to your destination and to ask for their advice. (Your own local centre will tell you the location of the nearest centre to your destination.) This is only really practical on a regional or (at most) national basis. Also, remember that it is quite tricky enough to forecast tomorrow's weather in your own locality – never mind the weather in two weeks' time for Japan where you may be going on vacation. The forecaster will not be holding back or keeping it a secret; at a local weather centre the density of plotted observation is high on the regional scale but is less on the nationwide scale and much less on the international scale. Information which is often available is the climatic average for various parts of the world at various times of the year. So really the best hope when going abroad is that the weather will behave in the way it usually does over a given location during a given season. This makes sense anyway, since many popular holiday locations have become so simply because they are known to have comfortable weather conditions at certain times of the year.

Longer range forecasting

So far we have been concerned with day to day forecasting. Basically, the method is to deduce the most recent weather trends from a study of present conditions. Detailed long range weather forecasts, that is for the next month or so, are not really within the capabilities of the amateur forecaster. The atmosphere 'knows' what its next move is even if we can't deduce it. After all, it is just a fluid that obeys the laws of physics. The trouble is that after a short interval of time the number of options open to it are so many and so varied that the present trends are not relevant. Clearly a different approach is necessary for the amateur working at home if he wishes to forecast the weather for the following month or season.

Various national meteorological services prepare long-range weather forecasts but these are – and should be regarded as – largely experimental. The different computer models used in the preparation of short-range forecasts are quite good at handling the slow movement and development of the upper air long waves (*p. 9*) for periods of up to about five days, and so can give good guidance on the overall weather expected for the next few days – a wet or a dry spell; a cold or a warm spell. However, the computers cannot cope for more than one or two days in advance with those migratory short waves that run through the long upper air waves and which steer the mobile anticyclones and depressions that determine most mid-latitude weather. If after only a few days' look into the future the computer forecast has started to go significantly wrong, then it is no use pursuing it further.

With long range forecasting one of the methods is simply to find another year like the present one. In principle, if an infinite number of years' weather had been catalogued, then this year's weather would have an identical 'twin' somewhere in that catalogue. So the idea is to take a summary of the past month's weather and then to go back through the years, matching this with the same calendar month year by year until an analogue – an exact copy – is found. If the two months match exactly then it is reasonable to suppose that the following months will also be similar. The forecast for the following month can then begin to take shape.

However, weather records covering the Northern Hemisphere go back only as far as 1873, so that the likelihood of finding an exact analogue is small indeed (records for the Southern Hemisphere are shorter). Also, the best representation of the motion of the atmosphere and the heat flow from equator to pole is the flow found at altitudes between 6000 metres and 12 000 metres (20 000–40 000 feet). These records only go back as far as 1945, although estimates of that flow can be made for years before 1945, working from the surface records, back to 1873.

The monthly mean pattern of surface temperature, surface pressure and upper air flow can also be catalogued and compared with this month's patterns. There appears to be some correlation between the upper air mean monthly flow and the rainfall distribution a month ahead, so this may prove to be a useful guide.

It is even more revealing to think in terms of anomalies. By taking an average over, say 30 years (generally accepted as being the shortest identifiable climatic period) any particular month can be described as warmer or cooler, or with higher or lower pressures than the average; patterns of anomalies in temperature and pressure can thus be drawn, and if the distribution of anomalies from years gone by is matched with the present month's anomalies, some indications for a long-range forecast begin to emerge.

Anomalies in surface conditions are also useful when matched up with averages. The extent of the arctic ice determines the extent of the pool of cold air over the pole; the amount of land covered by snow affects the radiation balance by reflecting sunlight; the sea surface temperature affects the temperature of the overlying air and its water vapour content. In fact the anomalies in the Atlantic Ocean surface temperatures go a long way towards revealing the surface pressure anomalies a month ahead which will affect the weather of Western Europe. As an example, during the winter of 1963 when the weather was so extreme over Great Britain, the British freeze was very long, lasting ten weeks until well into March. Although a ten-week freeze is not unusual in many areas of the world, even in parts of the continent of Europe, it is quite exceptional in Britain, and produced chaotic conditions, where bonfires had to be lit in the streets to prevent water pipes from freezing, and with massive power failures, forty-nine people died of cold. The most unusual thing about the weather at the start of the freeze was that the mean January pressure over Siberia was some eight millibars lower than normal, and off the Portuguese coast some six millibars lower than normal while over the Norwegian Sea pressure was some ten millibars higher than normal. In winters to come this pattern of surface pressure anomalies may well prove to be significant in heralding a further severe winter in Great Britain and Western Europe (all other things being equal).

A curious and rather interesting theory concerns the influence that the Moon may have on the weather. Just as the Moon is responsible for the ebb and flow of the ocean tides, so it causes tides within the atmosphere although with a monthly, rather than daily rhythm. This increases the probability that meteoric dust passing close to the Earth will be captured and make its way down to the troposphere. The abundance of condensation nuclei – the dust – assists cloud growth and therefore rainfall. According to this theory rain would occur more often on days following the maxima of the atmospheric tides. However, such a correlation has not yet been established.

In forecasting the weather for a whole season (three months, say) the technique of searching for analogues is unreliable, because the

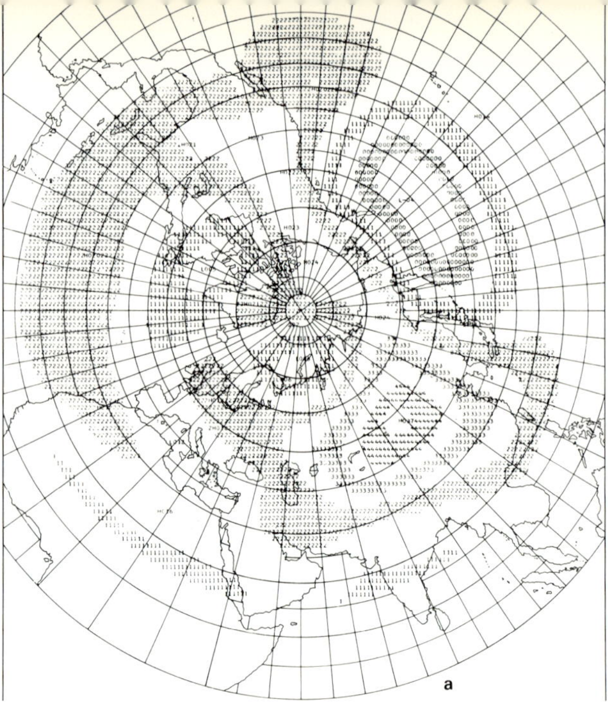

a

b

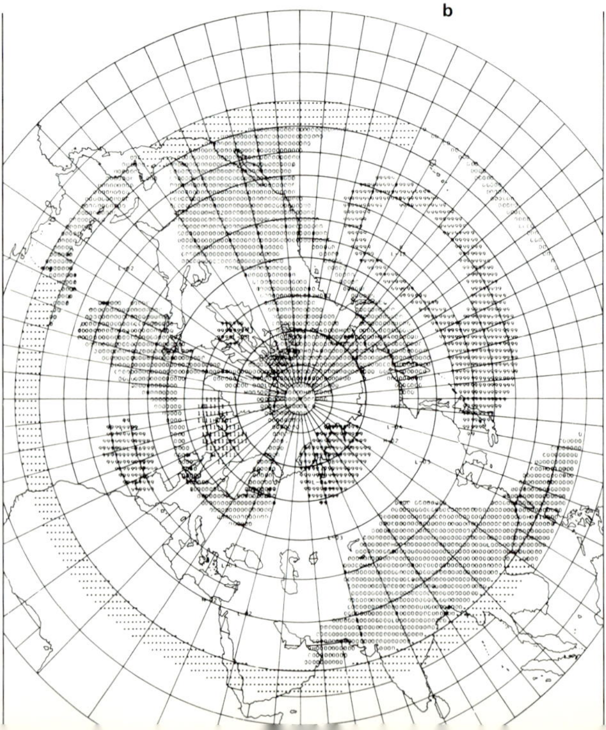

duration of the patterns which are to be matched is much shorter than the period of the forecast. An exception to this is the sea surface temperature for here anomalies may have a lifetime of three or four months and may therefore provide a basis on which to compare previous years' patterns. It will also be useful to try to compare the last three consecutive months' upper air and surface air anomaly patterns with those of the same period in previous years.

Usually, however, all that can be done is to resort to statistics and try to establish a correlation between certain weather patterns and the weather for the following season. Obviously this may be reasonably successful when dealing with an area which has a continental type climate, but is bound to be difficult to apply to those regions of the world such as British Columbia and Oregon, New Zealand, and Western Europe, which have maritime climates. Here there are no periods in the year which always have a specific type of weather.

Long-range forecasting even at its best is vague, and at its worst it is quite futile, suffering as it does from insufficient information in the first place. The accuracy sometimes turns out to be extremely good, and at other times very bad. Quantitative studies on the reliability of long-range forecasting have shown that it has 'some merit'. Progress in this field will, unfortunately, be slow – and that goes for all countries – because it takes time to build up a library of past years' weather patterns, and time to prove that a new technique is reliable.

The way to interpret any published long-range forecast is to realize that it is not intended to be a literal day-by-day guide to the next 30 (or 60 or 90) days' weather, but a general guide to the weather types expected to dominate over the period. The long-range forecast attempts to be accurate but not precise. For example 'below average' rainfall could result from every single day being drizzly, or from a few dry weeks being followed by rainy days in the last week, or from an alternation of dry days and wet days when the amounts of rain are always slight. Where and when it will rain within the forecast period cannot be specified until shortly beforehand. Precision is inherently low and everything mentioned is relative to the average for that time of year, because the forecast has to make its predictions in terms of anomalies or deviations of the monthly mean values from the climatic average. Rainfall may show a moderate deviation from the climatic average but temperatures may undergo a more extreme variation.

Fig. 59 A chart of the mean pressure for the winter of 1963 **a** where the various pressure ranges are indicated by different figures. Using these and data averaged over many years, the 1963 winter anomaly chart **b** was derived. This is discussed in the text.

Records and instruments

The average behaviour of the weather where you live can best be found out by checking past records for your own location. If you have not made any records it is worth considering doing so now. In years to come they will provide an invaluable source of information, and you may begin to notice definite patterns that will improve your forecasting skills.

Not everyone wants, or is able, to turn his garden into an observation station and few people have the time to prepare detailed accounts of each day's weather, but the basic essentials are readily observed and easily logged. In any case it is essential to be aware of the temperature, pressure and wind if you are going to formulate your own daily forecast, and even people in apartment blocks can measure pressure and temperature.

The first instrument to acquire is a barometer, or even better, although much more expensive, a **barograph**, which by means of a pen, continuously records pressure on a chart covering a week at a time. The instrument should be placed out of the sun and away from high temperatures, but can be indoors. Barometers should have a range of at least 960 mb or 72 cm of mercury to 1040 mb or 78 cm of mercury. Barographs usually have a rather greater range, typically 950 to 1050 mb. Before fixing in its permanent position – such as on a wall – the instrument must be 'set'. Telephone the nearest meteorological office and ask for the current value of sea-level pressure for your district. By means of the adjusting screw (located on the back of a barometer) set the needle or pen to the given figure. The instrument will now faithfully follow the ups and downs of atmospheric pressure and, in the case of the barograph, will record it continuously. When a barometer has been placed in position the knob on the front should be turned so that the pointer coincides with the needle showing the current pressure, and this should be done after any reading of the barometer. The position of the needle relative to the manually adjusted pointer will then indicate the trend since the last reading. Before reading a barometer it is a good idea to tap the glass gently to overcome the slight tendency of the needle to stick and thus not to show small pressure changes.

If a barometer is too expensive to buy it is quite easy to make an effective one at home (*Fig. 60*). Stretch some freezer film, or part of a balloon, over the opening of the largest diameter, rigid jar you can find, and use an elastic band to hold it tight over the opening. As the atmospheric pressure outside the sealed jar rises and falls, the membrane across the opening will be pressed inwards and outwards. The movement can be transferred to a card by means of a drinking straw. One end of the straw is glued to the centre of the membrane, the other end, with a needle stuck through it, points to a card with a scale which can be calibrated by borrowing a commercially made barometer.

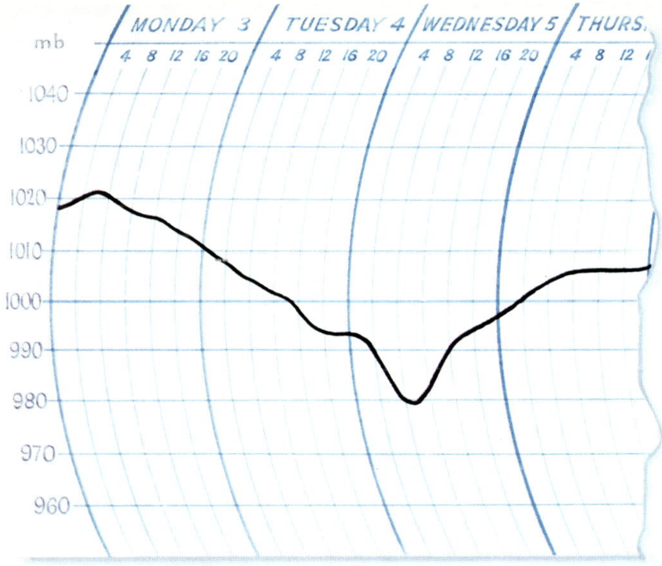

Fig. 60 Part of a chart produced by a barograph. Any depression with very rapid fall and recovery such as is shown here would be accompanied by strong winds and severe weather.

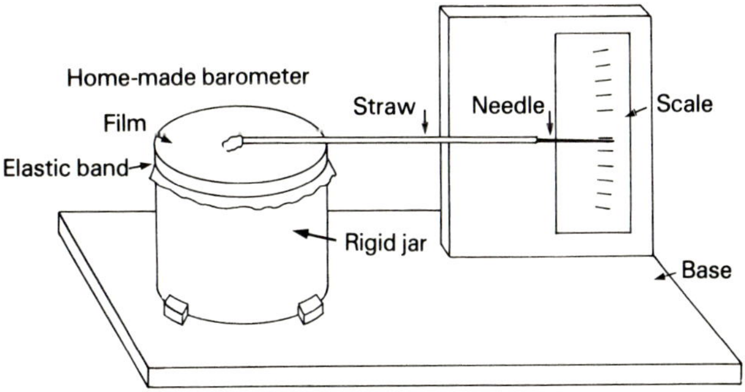

Fig. 61 A simple home-made barometer. The larger the jar, and the longer the straw the better. It may be necessary to try different films as some respond to changes in humidity or temperature.

Thermometers, however, are not expensive – so if possible, the best thing is to buy three, which will enable you to read the present temperature, the 24-hour maximum and minimum temperatures, and the humidity. Fix all thermometers about 1·2–1·5 metres (4–5 ft) above the ground on a north-facing wall so that they do not receive direct sunshine. The maximum and minimum thermometers are usually sold together as one instrument. Mercury, a very heavy liquid, expands as the temperature rises, pushing a metal index up the column. The index does not fall back down the column when the temperature drops, so that the position at which it is found indicates the **maximum temperature** since the last reading was taken. Alcohol, a low freezing-point liquid, contracts as the temperature falls and drags down a similar metal index to the lowest temperature. When the temperature rises the alcohol flows over the index without moving it so that its position indicates the **minimum temperature** since the reading was last taken. Each day the thermometers need to be reset by drawing the indexes back on to the meniscus (the surface of the mercury or alcohol) with the magnet provided. At any one time, simply reading the thermometer will give the actual **air temperature**.

The third thermometer is necessary to determine the **humidity**. This is done by wrapping the bulb of the thermometer in muslin cloth and leaving the end of the cloth in a small container of water fixed near to the bulb of the thermometer. This thermometer then shows the temperature at which water must be evaporated into the air in order to saturate it at that temperature, which is known as the **wet-bulb temperature**. By comparing this temperature with the actual air temperature the relative humidity of the air – the amount of water vapour actually in the air compared with the amount it would contain if saturated – can be read from tables (known as **hygrometric tables**) which you will need to buy – the local weather office will be able to tell you where you can get them. When the two temperatures are the same the air is 100% humid, and the greater the difference in the readings the drier the air. The humidity and temperature of the air reveal the predominant type of air mass and so give a clue to the weather which may be expected. Professional observers are, of course, equipped with more sophisticated instruments including thermographs and hygrographs for continuously recording temperature and humidity, but simple thermometers give good results.

Wind strength is measured professionally by **anemometers** or anemographs, and for anyone who likes tinkering with electrical gadgets it is fairly easy to make an anemometer from a small generator driven by a suitable shaft and cups, with a voltmeter used for reading windspeed after calibration. The more mechanically-minded might like to try a swinging plate anemometer as shown in *Fig. 63*, although if mounted on a high pole it may be difficult to read without binoculars. The greatest problem, with any wind measurements – and this also applies to simple wind-vanes – is in ensuring that they are not affected

The type of small thermometer screen known as the Bilham Screen, containing wet and dry bulb thermometers (vertical), and maximum and minimum thermometers (horizontal).

Fig. 62.

Maximum thermometer
(Diagrammatic)

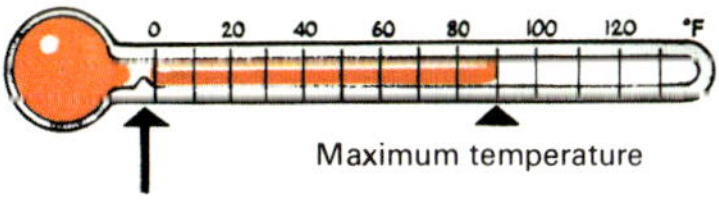

Mercury thread breaks at constriction as temperature falls. The thermometer is reset by shaking the mercury back into the bulb.

Minimum thermometer

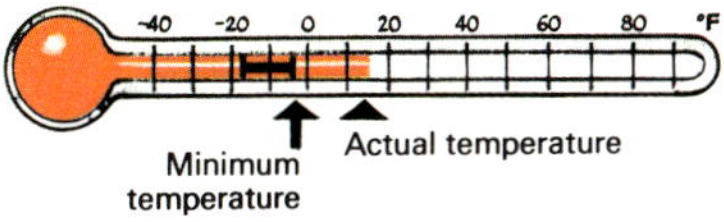

Index is drawn back to minimum position as temperature falls. The thermometer is reset by tilting it, bulb end upwards.

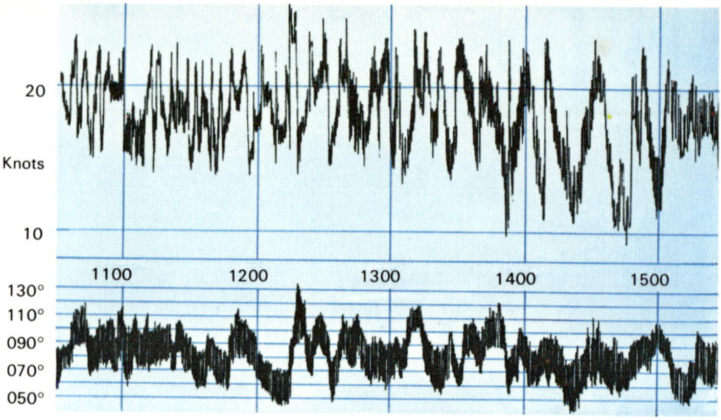
20
Knots
10
1100
1200
1300
1400
1500
130°
110°
090°
070°
050°

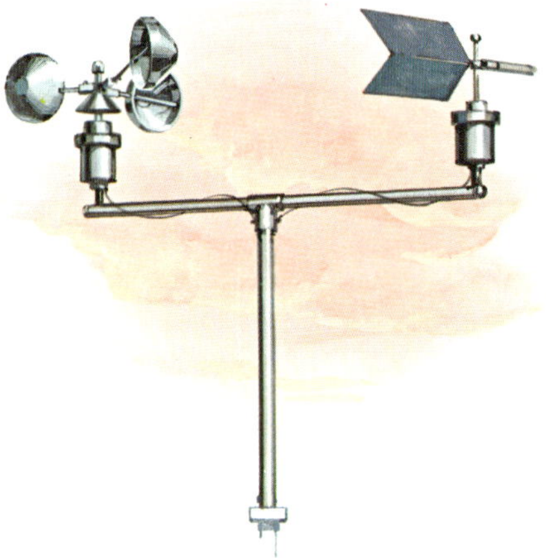

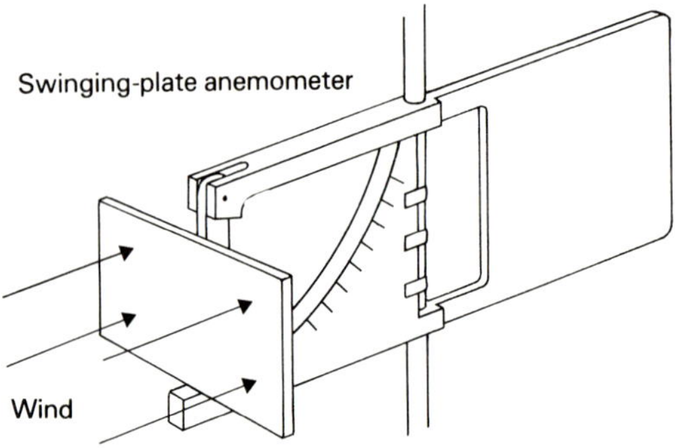
Swinging-plate anemometer
Wind

by nearby buildings, and professional instruments have to be mounted on high masts for this reason.

Despite these problems useful measurements of both direction and force can be made by improvising a wind sock like the ones to be found on airfields. An old bedsheet and two wire coat hangers bent into circles, one smaller than the other, can be sewn up into the basic sock shape and attached to a pole in a manner allowing it to swivel. (Kite materials would also be quite suitable.) The angle of the sock relative to the pole gives a reasonable indication of the wind speed; you can calibrate this approximately by holding the wind sock out of the window of a vehicle moving at a constant speed down a quiet road on a windless day. The wind direction can be gauged by using a compass, or better still, by marking the pole with the compass points. As a minimum, a single nail hammered in to indicate north would suffice. This will also aid in determining the wind direction at cloud level so that it can be seen from which direction the weather is coming. If constructing all this proves too much work or impractical, fair results can be obtained by simply attaching streamers to the top of a pole and guessing the wind speed and direction.

Fig. 63 *opposite centre* A generator-type cup anemometer and a remote-reading wind vane mounted on a single mast. These forms may be used with a suitable chart recorded to produce continuous charts of both wind speed and direction *opposite above*.

Opposite bottom A swinging-plate anemometer – a simple idea dating back to Leonardo da Vinci – which works surprisingly well once calibration has been carried out.

Fig. 64 Two forms of simple wind-sock, either of which will function well. It is essential that both types can swivel easily to face the wind.

Home-made wind socks

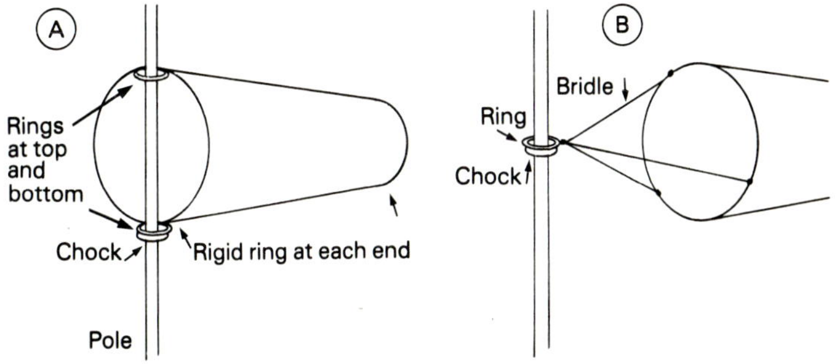

An airfield windsock must be positioned to be clear of all possible ground eddies, but at about 10 m (33 ft) the air flow will normally be fairly steady.

All these weather observations should be made at the same time each morning and recorded simply in a log book under the headings of 'pressure', 'air temperature', 'wet-bulb temperature', 'humidity', 'wind' (direction and strength), 'maximum temperature' (which will normally be for the previous day) and 'minimum temperature' (which normally occurs during the early morning around dawn). At the very least it is important to log the maximum and minimum temperatures: these will prove extremely useful in aiding the forecast for the same time in succeeding years.

Two measurements have not yet been mentioned, those of sunshine and of rainfall. Measurement of the duration or intensity of sunshine requires special instruments and is thus not possible for the amateur. The old-fashioned black-bulb thermometers are sometimes seen, passed down from a previous generation of amateur meteorologists, but are unreliable as two rarely give the same reading and the interpretation of even a single instrument's results is uncertain, so they remain more of curiosity, rather than scientific, value. Professionally reporting stations, not equipped for sunshine recording, restrict themselves to describing cloud cover by means of symbols.

It might at first be thought that rainfall measurement would be easy, but this is not the case, and careful attention has to be given to the

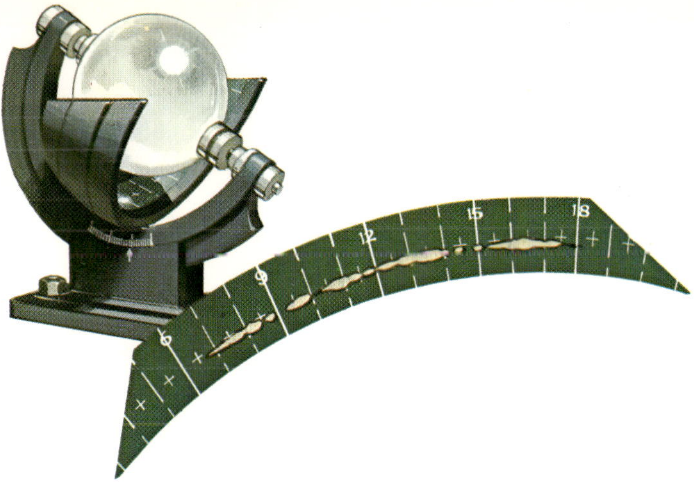

Fig. 65 One form (the Campbell-Stokes) of sunshine recorder. The glass sphere focuses the rays from the Sun onto a card which is charted to give a record of sunshine hours.

construction, and particularly to the siting, of suitable instruments. The collecting funnel must be sharp-edged and of a precisely known area, the shape is such as to minimize rain splashing out again, and the position of the whole gauge has to be arranged so that eddies and splashing from the ground are eliminated as far as possible. A general description of rainfall may of course be made by use of the standard symbols, as may any of the other special phenomena.

It will be of interest to try to record the cloud types present and this may be done in a general way from the information and illustrations given here. Anyone wishing to record full details is recommended to obtain one of the cloud classification publications noted in the bibliography, which give the official coding methods for cloud types and amounts. Any of the information recorded in the ways described above may be preserved in the form of what may be called a 'weather diary', which will be of considerable interest and use. Anyone who wishes to progress from this, and to report observations to their national service will naturally have to make sure their equipment meets the requirements – such as mounting thermometers in a Stevenson screen – and record and report observations in specific ways. However many persons do just this, both for synoptic (i.e. forecasting), and in addition for climatological purposes.

Forecasting likely maximum temperatures

Even if you are not compiling your own weather records it is of very considerable practical value to have information on maximum and minimum air temperatures for your own locality. These figures for your own area should be readily available from a local library or weather centre, possibly in the form of values for ten-day periods throughout the year. A typical extract from such a listing is shown in the table below.

To forecast the likely maximum temperature it is necessary to have such a table and a means of obtaining the maximum air temperature. From this reading the average for the time of year should be subtracted, to give a measure of the relative warmth, that is from 'hot for the time of year' to 'cold for the time of year'. A consideration of the origin of the air (polar or tropical), sky conditions, and strength of wind will enable a judgment to be made of the day's likely temperature – again 'for the time of year'.

The identification of the air stream may cause some slight difficulty, unless it is already known from other information, but for practical purposes may be judged quite well from the 'feeling in the air'. The dry, extreme cold of continental polar (or arctic) can be distinguished from the penetrating damp cold of maritime arctic air, while maritime polar air – although damp – tends to have a more exhilarating feel. The humidity of maritime tropical air means that it can be easily told from the dry heat of continental tropical air. Just going outside to judge the quality of the air, and comparing the impression gained with the officially reported air mass will soon enable you to estimate the likely air stream. Even the simple rule that polar air in the northern hemisphere will approach from directions between west through north to east (270°–360°/0°–90°) and tropical air from south-west through south to south-east (225°–180°–135°) will give an approximate idea of the origin, although obviously allowance should be made for regional effects.

Period (approx. 10 days)	Max. Temp. °C	Min. Temp. °C
March 2–11	9	2
12–21	10	2
22–31	11	4
April 1–10	12	4
11–20	13	5
21–30	14	6
May 1–10	15	7

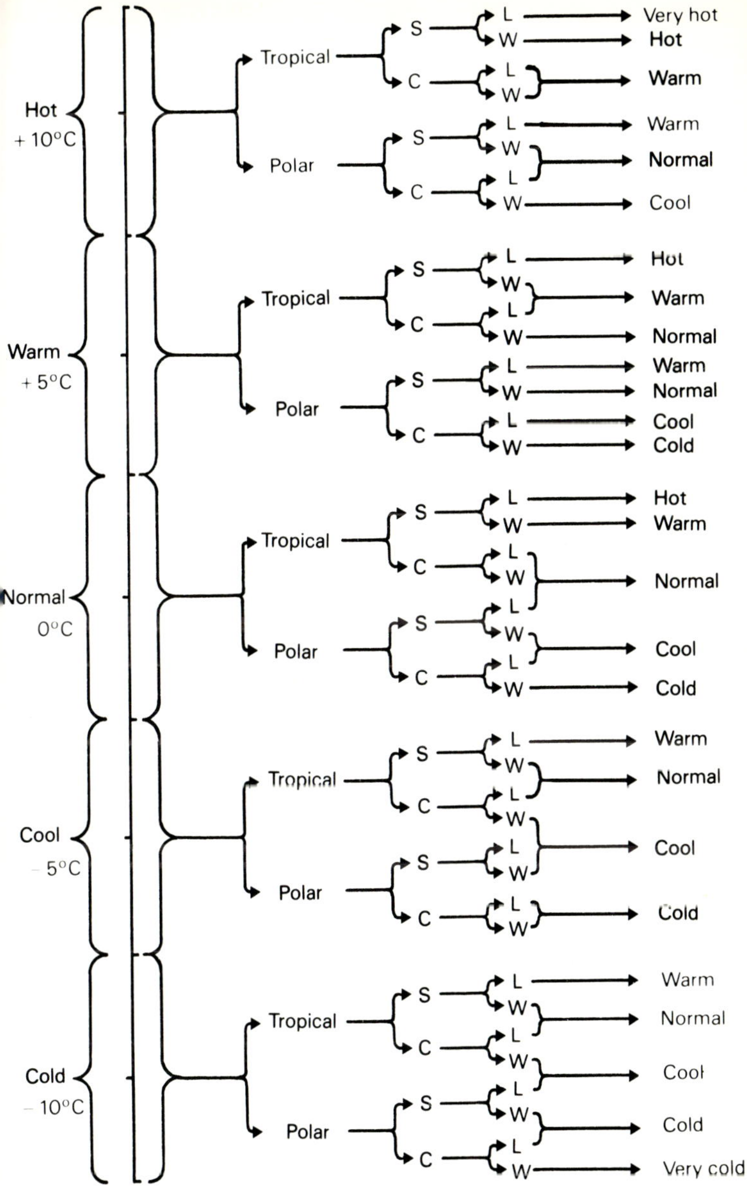

Fig. 66 Graphic forecast diagram for determining likely maximum temperature. C=Overcast, cloudy; S=Some sunshine; W=Windy; L=Light winds only.

Forecasting overnight temperatures and precipitation

There are various methods of forecasting overnight temperatures, particularly for judging the occurrence of frost, which involve knowledge of wet and dry bulb temperatures in the early afternoon (at 3 pm), wind speed and, in some methods, cloud cover and days since rainfall. However if conditions are expected to remain clear overnight a simplified method will give quite good results. For this it is necessary to know the time of sunset and of dawn, and to be able to take the air temperature. After reaching its afternoon peak the air temperature begins to decline, and a short time after sunset adopts a constant rate of fall as the Earth's heat is radiated away to space. The method therefore consists of taking two air temperature measurements an hour apart, with the first a certain time after sunset. In theory the interval between sunset and the first measurement should be governed by latitude and month, the interval after sunset being less at lower latitudes. However the following will work satisfactorily for most temperature latitudes.

Northern Hemisphere	Interval between sunset and first measurement	Southern Hemisphere
December, January	1 hour	May, June
October, November, February	$1\frac{1}{2}$ hours	March, April, July
March, April	2 hours	August, September

When the rate of fall becomes known from the second air temperature measurement, it can be seen if the temperature will reach 0°C (32°F) in the night. The lowest temperature will be reached at about the hour before dawn, assuming of course, that the sky remains fairly clear.

A method of forecasting likely precipitation is shown in *Fig. 68*, and this is largely self-explanatory. It should, of course, be used in conjunction with the individual cloud descriptions and the details of depressions and shower/thunderstorm sequences given earlier.

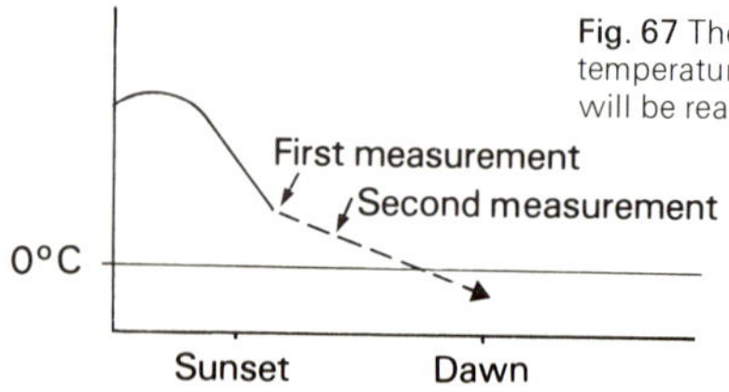

Fig. 67 The method of forecasting overnight temperatures (see text). Lowest temperature will be reached just before dawn. See also *Fig. 35*.

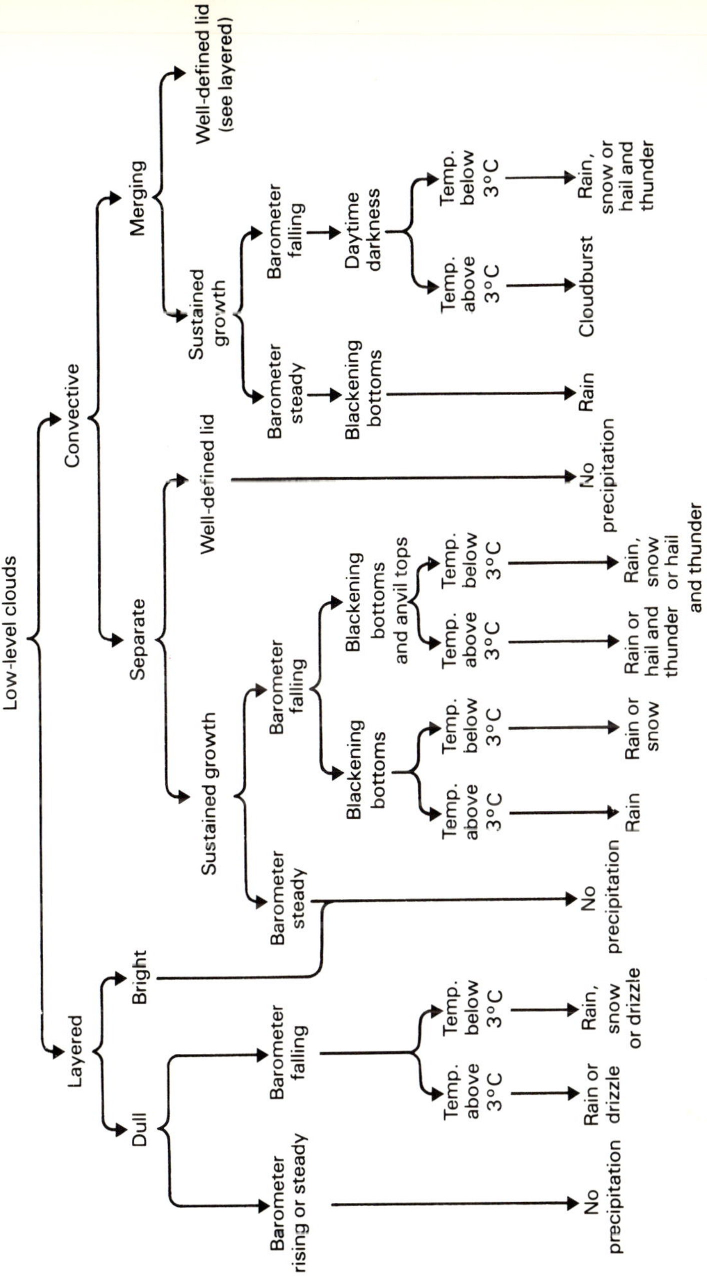

Fig. 68 Graphic forecast diagram for determining likely precipitation.

Making a forecast

It may be helpful to outline the factors which should be considered in making a forecast, as well as giving an indication of where these are discussed in greater detail.

The state of the sky and the clouds
Clouds give very important indications of both the weather existing at the time, as well as early clues to impending changes. The various forms of cloud and optical phenomena are described on *pp. 30-59*, but more general details of conditions under various pressure systems are also relevant (*pp. 24-27*).

Winds
Wind direction at various heights may be assessed by cloud movement, and the relative motions are important in determining the likely weather around an approaching low pressure system (*pp. 88-97*).

Pressure and pressure changes
The approach of a depression (*p. 88*) is heralded by falling pressure, while strong winds accompany steep rises and falls (*p. 18*). A fall and subsequent steadying of pressure, or a fall and rise, will occur with the passage of a front. High pressures and slow changes show anticyclonic activity (*pp. 24-27*).

Air mass
The predominant and encroaching air masses (*p. 20*) will closely determine the weather pattern, although frequently overridden by the influence of pressure, most especially under anticyclonic conditions.

Temperature
The development of daytime temperatures will closely control cloud formation (*p. 30*) which may lead to showers (*p. 98*) or, with high temperatures, thunderstorms (*p. 102*). The fall in night-time temperatures, in particular, will determine the likelihood of the formation of fogs and frost (*pp. 74-81*).

Humidity
High relative humidity and low temperatures will indicate the possibility of fog, dew or frost (*pp. 74-81*). High humidity and high temperatures are necessary for the formation of thunderstorms (*p. 102*). The approach of a front is heralded by increasing humidity, while low humidity generally accompanies anticyclonic conditions (*p. 24*).

Local conditions
Great alterations from the area conditions can occur when local factors intervene. This particularly applies to cloud amounts (*p. 30*), rainfall (*p. 46*) and the formation of fogs and frosts (*pp. 74-81*).

Glossary of terms

Adiabatic 'Without the addition or loss of heat', as in a mass of air no longer in contact with the ground.
Aerosol Minute solid or liquid particles suspended in the atmosphere.
Ambient Relating to the surrounding air. A rising thermal will have a temperature above the ambient temperature.
Beaufort scale Wind speed estimated on a numerical scale, ranging from 0, calm; 1, 1-3 knots (0·3-1·5 m/s or about 1-3 mph) to 12, above 64 knots (above 33 m/s or above 73 mph).
Celsius The correct term for the temperature scale – frequently, and incorrectly, called 'Centigrade' – where the freezing and boiling points of water are 0°C and 100°C.
Continental climate A climate typical of continental interiors, with large temperature ranges and low rainfall.
Convergence In simplified terms, the accumulation of air within a given volume of the atmosphere. It is the opposite of divergence.
Divergence In simplified terms, the depletion of the amount of air within a given volume. Like convergence, it also has a more specific, technical meaning in meteorology.
Hadley cell Simple thermal circulation between latitudes 0° and 30° with a high-level poleward flow from the heated regions.
Kelvin A unit of heat (K) used to form a temperature scale beginning at absolute zero (−273·15°C). 0°C=273·15 K.
Lapse rate The decrease of temperature for a unit increase in height. Important rates are the dry adiabatic (q.v.) lapse rate, the saturated lapse rate and the environmental lapse rate, the latter being that actually pertaining in the atmosphere at any particular time.
Latent heat The quantity of heat absorbed or emitted when a substance changes state without any change of temperature, for example ice to water, ice to water vapour or the reverse reactions.
Macroclimate The climate of a large region of the world.
Microclimate The climate of a very small area; frequently used to describe the effects upon plants and insects.
Maritime climate A climate dominated by the nearness to the sea, and usually having small temperature ranges and fairly moist conditions.
Relative humidity The amount of moisture in the air, usually given as a percentage of the amount which the air would contain when fully saturated at a given temperature.
Synoptic Relating to a general view at a particular time. A synoptic chart shows standard data for a set observational time.
Stability A measure of the way in which a parcel of air will act when it is displaced vertically. Under stable conditions it will seek to regain its original level; with unstable conditions it will continue its motion.
Wind-chill The loss of heat by the skin purely due to the effects of wind. Even a moderate wind will produce as much heat loss as will occur in low temperatures and calm conditions.

Bibliography

Many excellent government publications exist and the following can be fully recommended:

National Oceanic and Atmospheric Administration (NOAA) publications – obtainable from U.S. Government Printing Office, Washington, D.C. 20402 (request Subject Bibliography SB-234, *Weather*):

Cloud Code Chart, 1972
Clouds, 1974
Spotter's Guide for Identifying and Reporting Severe Local Storms, 1975
Storm Surge and Hurricane Safety, 1978
Tornado, 1978
Weather for Aircrews, 1974 (Specialized)

United Kingdom Meteorological Office publications – obtainable from H.M. Stationery Office, London (request Sectional List 37, *Meteorological Office*):

Cloud Types for Observers, 1962
Elementary Meteorology, 2nd edn., 1978 (*Not* for beginners)
Handbook of Aviation Meteorology, 2nd edn., 1971 (Specialized)
Meteorological Glossary, 1972
Meteorology for Mariners, 3rd edn., 1978 (Specialized)
Observer's Handbook, 4th edn. in press

Non-government publications at various levels:

Barry, R. G. and Chorley, R. J., *Atmosphere, Weather & Climate*, 3rd edn., Methuen, London, 1976
Battan, L. J., *Weather*, Prentice-Hall, Englewood Cliffs, N.J.
Cole, F., *Introduction to Meteorology*, 2nd edn., Wiley, New York, 1975
Crawford, J., *Mariner's Weather*, Nortons, New York, 1978
Holford, I., *Guiness Book of Weather Facts and Feats*, Guiness, London, 1977
Pedgley, D., *Mountain Weather*, Cicerone, Cumbria, 1979. (Obtainable from Royal Meteorological Society – address given below.)
Scorer, R. S., *Clouds of the World*, David & Charles, Newton Abbot, 1972
Scorer, R. S. and Wexler, H., *A Colour Guide to Clouds*, Pergamon, Oxford, 1963. (Obtainable from Royal Meteorological Society – address given below.)
Wallington, C. E., *Meteorology for Glider Pilots*, Murray, London, 1977
Watts, A., *Weather Forecasting, ashore and afloat*, Adlard Coles, London, 1967
Watts, A., *Instant Weather Forecasting*, Adlard Coles, London, 1968
Watts, A., *Wind and Sailing Boats*, David & Charles, Newton Abbot, 1973

Periodicals:

Weather, monthly, published by Royal Meteorological Society, James Glaisher House, Grenville Place, Bracknell, Berks. RG12 1BX
Weatherwise, bi-monthly, published by American Meteorological Society, Boston.

Surface weather map and station weather

Explanation of symbols

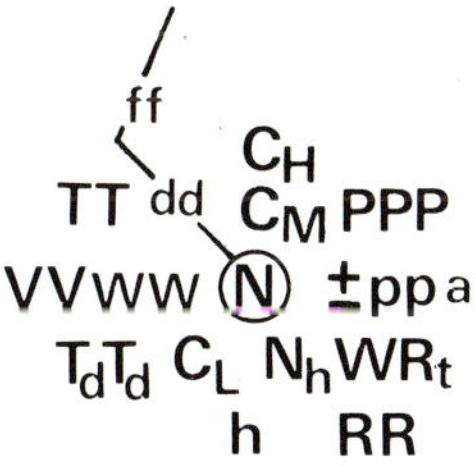

MODEL

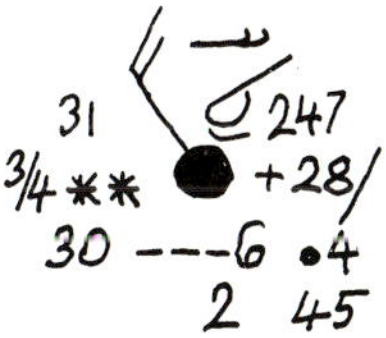

REPORT

N Total amount of cloud 8=completely covered

dd True direction from which wind is blowing 32=320°=NW

ff Wind speed in knots 20=20 knots

VV Visibility in miles and fractions $8=\frac{8}{16}$ or $\frac{1}{2}$ mile

ww Present weather 71=continuous slight snow

PPP Barometric Pressure (in millibars) reduced to sea-level 247=1024·7 mb

TT Current air temperature

W Past weather 6=rain

C_L Cloud type 7=Fractostratus and/or Fractocumulus of bad weather (scud)

h Height of base of cloud 2=300–599 feet

C_M Cloud type 9=Altocumulus of chaotic sky

C_H Cloud type 2=Dense cirrus in patches

T_dT_d Temperature of dewpoint

a Characteristic of barograph trace 2=rising steadily or unsteadily

pp Pressure change in 3 hours preceding observation 28–2·8 millibars

Sky Coverage

N USA/*Elsewhere*

 No clouds/*No clouds*

 One-tenth or less/*One-eighth*

 Two-tenths or three-tenths/ *Two-eighths*

 Four-tenths/*Three-eighths*

 Five-tenths/*Four-eighths*

 Six-tenths/*Five-eighths*

 Seven-tenths or eight-tenths/*Six-eighths*

 Nine-tenths or overcast with openings/*Seven-eighths*

 Overcast/*Overcast*

 Sky obscured/*Sky obscured*

Cloud types

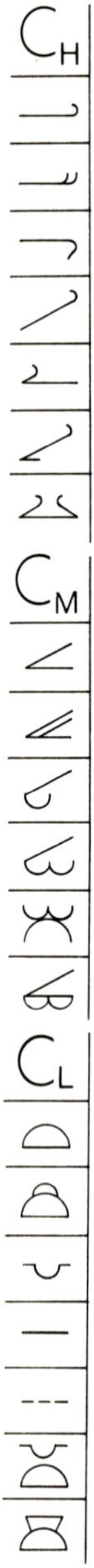

C_H	HIGH CLOUDS
	Filaments of Ci, or 'mares tails' scattered and not increasing
	Dense Ci in patches or twisted sheaves, usually not increasing, sometimes like remains of Cb; or towers or tufts
	Dense Ci, often anvil-shaped, derived from or associated with Cb
	Ci, often hook-shaped, gradually spreading over the sky and usually thickening as a whole
	Ci and Cs, often in converging bands, or Cs alone; generally overspreading and growing denser; the continuous layer not reaching 45° altitude
	Ci and Cs, often in converging bands, or Cs alone; generally overspreading and growing denser; the continuous layer exceeding 45° altitude
	Veil of Cs covering the entire sky
C_M	MEDIUM CLOUDS
	Thin As (most of cloudlayer semi-transparent)
	Thick As, greater part sufficiently dense to hide sun (or moon), or Ns
	Thin Ac in patches; cloud elements continually changing and/or occurring at more than one level
	Thin Ac in bands or in a layer gradually spreading over sky and usually thickening as a whole
	Ac formed by the spreading out of Cu
	Double-layered Ac, or a thick layer of Ac, not increasing; or Ac with As and/or Ns
C_L	LOW CLOUDS
	Cu of fair weather, litter vertical development and seemingly flattened
	Cu of considerable development, generally towering, with or without other Cu or Sc, bases all at same level
	Sc not formed by spreading out of Cu
	St or Fs or both, but no Fs of bad weather
	Fs and/or Fc of bad weather (scud)
	Cu and Sc (not formed by spreading out of Cu) with bases at different levels
	Cb having a clearly fibrous (cirriform) top, often anvil-shaped, with or without Cu, Sc, St, or scud

Wind speed

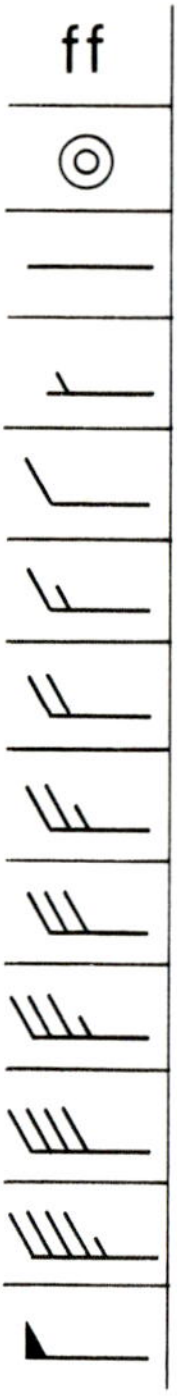

ff	Knots
	Calm
	1–2
	3–7
	8–12
	13–17
	18–22
	23–27
	28–32
	33–37
	38–42
	43–47
	48–52

WW Present Weather

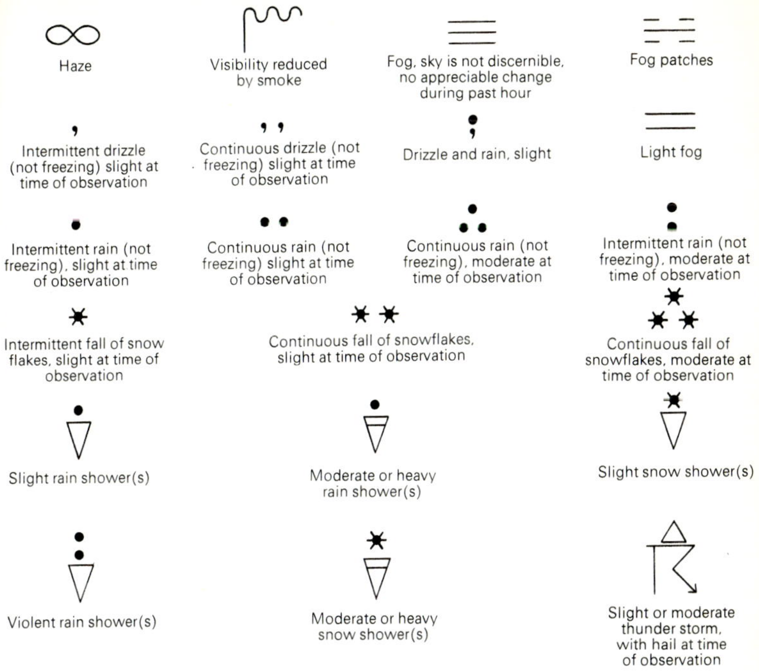

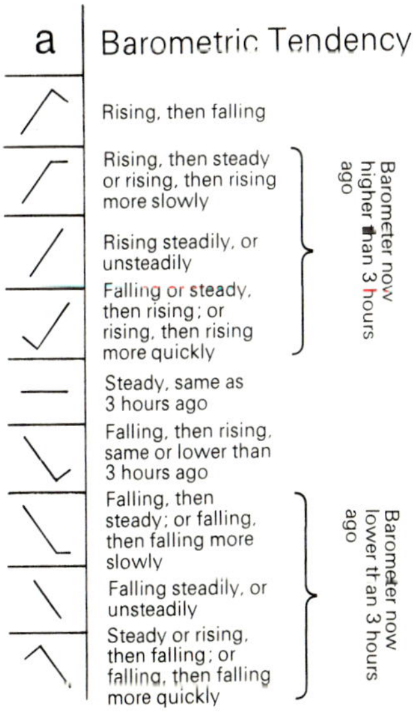

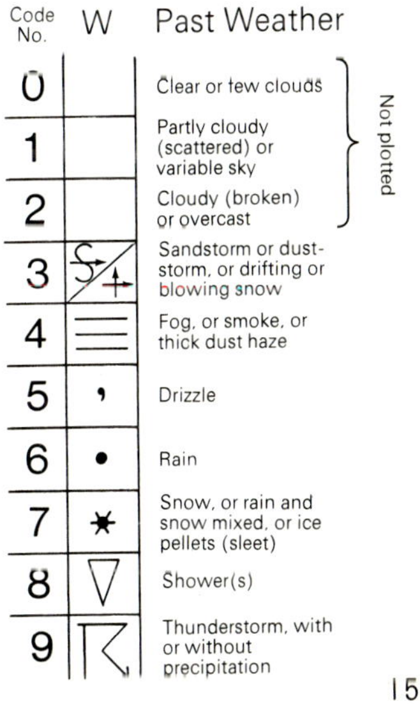

Index

Numbers in *italics* refer to illustrations. The glossary has not been indexed.

adiabatic lapse rate 32
Adriatic 114
advection fog 74, *75*
Africa *6*, *28*, *36*, *55*, *100*, *104*, *107*
air frost 66, 78, 80
air mass 9, *12*, 13, *13*, 16, 20–23, *20*, 26, *30*, *112*, *114*, *116*
air temperature 150
air wave 60–61
Alps *51*, *62*, 66, *82*, *94*, 114
altocumulus 36, 42, 43, *43*, 45, *47*, *55*, 56, *57*, 93
altocumulus castellanus 43, *43*
altostratus 36, 42, *42*, 44, 48, *90*
anabatic wind 68
ana cold/warm front 27, 88, *92*
analogues 136
anemograph 142
anemometer 142, *145*
anomalies 137, 139, *139*
Antarctic Circle *6*, 69, 116
anticyclone 17, 24, 96, 114, 136
anvil cloud *36*, 38, *103*
Arabia *6*
arctic air 20, 22, 112
Arctic/Antarctic Front 24, 112
Arctic Circle 11, 20, *21*
arctic sea smoke 76, *77*
Arizona 106, 112
Asia 12, 13
Atlantic 114, 118, *127*, *129*, 137
atmosphere 6–9
atmospheric layers 28–29
atmospheric pressure 14–17, 22, 40, 105, 122
atmospheric temperature 28–29, *29*
Australia 11, 23, 111, 116, *116*

barograph 140
barometer 9, 14, 15, 40, 75, 121, 140, *141*
Bering Sea *21*
Bilham Screen *143*
black ice 80
Black Sea 114
blocking high 24
blue moon 53
Bora 114
Brickfielder 116
British Columbia 139
Buys-Ballot law 18, 122

California 86
Canada *23*, 112
Canary Islands *58*
Cape Verde 111
carbon dioxide 84
China Sea 111
Chinook wind 65, 112
circulation of atmosphere 6–9, *7*, 12
cirriform cloud 54
cirrocumulus 44, 45, *45*
cirrostratus 42, 44, *44*, *55*, *90*, 93
cirrus 19, 36, 45, *48*, 88, *89*, 93, *94*, 109
cloudburst 64
cloud colour 50–53, 56
clouds 15, 16, 17, 19, 28, 29, 30–53, *30*, *33*, 54, 62, 64, 90, 92, 129, 147
coalescence 46
coastal weather *16*, 70–73, 82
cold front 26, 27, 92, *92*, 93, 96, *97*, 105, 116, *122*, *127*
Colorado 112
computers 136
condensation level *34*, *35*, *39*, *41*
continental air 20
continental arctic air 148
continental polar air 20, 22, 112, 114, 148
continental tropical air 20, 23, 114, 116, 148
convection *8*, *14*, 98
convective cloud 46
converging air *8*, 9, 15, 122
Coriolis effect 18
corona 42, 54, *55*
cumiliform cloud 32
cumulonimbus 30, 32, 34–36, *33*, *36*, 40, 48, 92, 93, *97*, *100*, 101, 105, 108, 109
cumulus *14*, *17*, 32, 34, *35*, 36, 38, 40, 43, *48*, 63, 72, 88, 93, 98
cyclone 17, 23, 111

depressions *6*, 17, *17*, 18, 24, *24*, *26*, 27, 28, 88–97, *89*, *94*, *97*, 112, 114, *123*, *127*, 129, 136
dew 46, 56
dewpoint 30, 32, *39*
diamond dust 76
diverging air *8*, 9, 15
drizzle 46, 48, 72, 96
dry adiabatic lapse rate 32
dust whirl 106

easterlies 7, 23
equatorial air 20, 22, 116
Eurasia 20, 22
Europe *71*, 114, *114*, *131*
extra-tropical cyclone 24
eye of storm 109

fall-streak 45, *47*
fall wind 68, 69, *69*
fog 46, 54, 56, 58, 68, 74–77, *77*, 79, 84, 114
fogbow 56, *56*, 58
föhn wind 65, *65*, 112, 114
forecasting terms 120
France *69*, *94*
freezing fog 48, 54, 76, 79
freezing rain 48, 80, 90
fronts *20*, 23, *23*, *24*, 26, *30*, 44, *89*, 92,

93, *94*, 96, *97*, 105, 116, 122, *122*, 123, 127
frost 46, 68, 78–81, 84, 135, 150

geostationary satellite 127
glazed frost 80
gliding 34, 60, 61
glory 56, *56*
Great Britain *61*, *64*, *87*, *118*, *128*, 137
Great Lakes 112
Great Plains 112
Greenland 78
ground frost 78
ground ice 80
Gulf of Mexico 106, 111, 112
Gulf Stream *23*

hail 46, 98, *100*, 101
halo 44, 54, *55*, 88, 90
hay fever 82
haze 50, 76
heat island 84
heiligenschein 56
high-altitude ice-clouds 54
high clouds 32, 42, 46
high pressure systems (highs) *8*, 16, 17, 18, 105, 114, 122, 123
hill fog 72, 74
hoar frost 78, *79*, 80
horizontal convergence 42
humidity 142, 146, 148
hurricane 109, *110*, 111, 112
hygrometric table 142

ice 80, 101, 102
ice cloud 43
ice crystal 29, 36, *36*, 45, 46, 54, 76
ice fog 76
Iceland 114
India 13, *85*
inversion 24, 32, *39*, 40, *41*, 86
iridescence 56, *57*
isobar 18, *18*, *24*, 122, 123
isobaric chart 18, 121, *121*, 122, 132, 134

jet stream *8*, 18, 19, *19*, 45

katabatic wind 68, *69*, 114
kata front 27, 96, *97*
Kenya *55*, *104*, *107*
Khamsin 114
Kilimanjaro *28*

Labrador Current 112
lapse rate 29, 32, 102
layer cloud 32, 38, 42, 46, 48, 49, 80
lead pollution 87
lee-waves 28, 60, *61*
lightning 102, *105*
long range forecasting 136–39
long waves 9
low cloud 32, 46
low pressure systems (lows) *8*, *15*, 16, *16*, 17, 18, *18*, 19, 24, 26, 32, 40, 48, 88, 91, 92, 98, 105, 114, 116, 122, 123

maritime air 20
maritime arctic air 20, 22, 148
maritime polar air 20, 22, 23, 114, 116, 148
maritime tropical air 20, 23, 112, 114, 116, 148
maximum/minimum temperature 134, 142, 146, 148
Mediterranean 114, 116
Mediterranean Front 114
Meteorological Office 118
middle cloud 19, 32, 42
millibar 14
Mississippi 112
mist 74
Mistral 66, 114
mock sun 54, *55*, *73*
monsoon 116
monsoon wind 13
Moon 137
mountain-climbing 68
mountains *28*, *30*, 60–61, 62, 68

nacreous cloud 28
National Weather Service 118
Newfoundland 112
newspaper forecast 134
New Zealand 23, 116, *116*, 118, 139
night fog 66
nimbostratus 36, 42, 48, 49, 90, 92
noctilucent cloud 28–29
North Africa 23
North America 12, 20, 22, 23, 112, *112*
North Pole 111

occlusion *24*, 26, 27, *93*
occluded front 26, 93, *93*, *127*
orographic cloud *30*, 62, *63*, 64, *64*
orographic uplift 105
Oregon 139
overnight temperatures 150–51
ozone 86

Pacific 118
pea souper 85
peroxy-acetyl nitrate 86
photochemical smog 86, 87
polar air 20, 22, 23, *23*, 24, 112
Polar Front 23, 24, 112
polar regions 49
pollen 82
pollution 24, 75–76, 84–87, *85*, *86*, *87*
precipitation 42, 46–49, *47*, *48*, 90, *100*, 105, 130, 150–51, *151*
pressure 7, 9, 12, 136, *139*
professional forecasting 118

radar *118*, 121, 130, 135
radiation fog 74, *75*
radio forecast 132, 134

rain *15*, 16, *17*, 42, 46, 48, *48*, 49, 62, 88, 90, 92, 96, 98, 101, 116, *123*, 139, 146, 147
rainbow 58, *58*, *59*
red sky at morning/night 53
Rhône valley 66
ridge 122, 123
rime 76, 79, *81*
Roaring Forties 116
Rockies 66, 112
Russia 114

sand devil 106
satellite imagery *6*, *71*, *72*, *110*, 126–29, *126*, *127*, *128*, *131*, 132
saturated adiabatic lapse rate 32
Scandinavia 114
sea breeze 70, *70*
sea fog *73*, 76
seasonal effects 10–13
sea temperature 70, 137
secondary waves 9
settled weather 9, 17, 24
short waves 136
shower cloud 34
showers *17*, 36, 63, 98
Siberia 12, 137
Sirocco 114
sky colour 50–53, 54
sleet 46
smog 24, 75, 85, 86
smoke fog 85
snow *28*, 42, 46, 49, 62, 90, 114
snowflakes 46, *48*
snow line 64
soil 82
Southerly Buster 116
squall line 105, 106
stable air 32
steam fog 76
stratiform cloud 32
stratocumulus 24, *26*, 38–41, *38*, *39*, *40*, *41*, 44, 61, 88, *94*, 96, *97*
stratosphere 28, 34, 101
stratus 46, *47*, *63*, 96, *97*
sulphur dioxide 75, 84, 85, 86
sun colour *50*, 52, 53, *53*
sunshine measurement 146, *147*
sunslope 63
supercooled water droplets 30, 36, 42, 46, 101
surface friction 18
surface temperature *10*, 136
Switzerland *47*, *62*, *82*

Tasmania 116
telephone forecast 134
television forecast 121, 132, 134
thermal 34, *34*, *39*, 40, *40*, 41, 82, 84, 86, 98
thermometer 142, 147
thunderstorm 36, *97*, 98, 102–105, *102*, *103*, *104*, 106
tornado 106
tropical air 20, 22
tropical cyclone *108*, *109*, 111, 116
tropical storm 108–11
tropopause 28, 29, *36*
troposphere 28, 29, *29*, 32, 49, 137
trough 122, 123
turbulence 60, 61
typhoon *110*, 111

unsettled weather 9
unstable air 32, *43*
upper air flow 137
upper atmosphere 9
upper atmosphere waves *20*, 23, 112, 136
upslope fog 74
urban weather 84
U.S.A. *21*, *23*, *48*, *53*, *71*, *72*, *75*, *102*, 106, 111, 112, *112*, *133*

valley fog *69*, 75
valleys 66–69, *66*, *67*, *68*, *69*, *82*
vegetation 82, *82*
Vesuvius *29*

warm air mass 42
warm front *23*, 45, 54, *89*, 91–92, 93, *94*, 96, 105, *127*
water devil 106
waterspout 106
West Coast 112
West Indies 111
westerlies 7, 23, 112
Western Europe 22, 40, 137, 139
wet-bulb temperature 142
whirlwind 106, *107*
willy-willie 111
wind *16*, 18–19, *19*, 41, 60–61, 84, 98, 101
wind chill factor 80
wind direction 18, 19
wind sheer 23
wind sock 145, *145*, *146*
wind speed 145, 146
wind-vane 142
wind veer 90, 93